Lecture Notes in Artificial Intelligence   5669

Edited by R. Goebel, J. Siekmann, and W. Wahlster

Subseries of Lecture Notes in Computer Science

Thanaruk Theeramunkong   Cholwich Nattee
Paulo J.L. Adeodato   Nitesh Chawla
Peter Christen   Philippe Lenca
Josiah Poon   Graham Williams (Eds.)

New Frontiers in Applied Data Mining

PAKDD 2009 International Workshops
Bangkok, Thailand, April 27-30, 2009
Revised Selected Papers

 Springer

Series Editors

Randy Goebel, University of Alberta, Edmonton, Canada
Jörg Siekmann, University of Saarland, Saarbrücken, Germany
Wolfgang Wahlster, DFKI and University of Saarland, Saarbrücken, Germany

Volume Editors

Thanaruk Theeramunkong, E-mail: thanaruk@siit.tu.ac.th
Cholwich Nattee, E-mail: cholwich@siit.tu.ac.th
Paulo J.L. Adeodato, E-mail: pjla@cin.ufpe.br
Nitesh V. Chawla, E-mail: nchawla@nd.edu
Peter Christen, E-mail: peter.christen@anu.edu.au
Philippe Lenca, E-mail: philippe.lenca@telecom-bretagne.eu
Josiah Poon, E-mail: josiah@it.usyd.edu.au
Graham Williams, E-mail: graham.williams@togaware.com

Library of Congress Control Number: 2010931089

CR Subject Classification (1998): I.2, H.3, H.4, H.2.8, J.1, I.4

LNCS Sublibrary: SL 7 – Artificial Intelligence

ISSN   0302-9743

ISBN-10   3-642-14639-2 Springer Berlin Heidelberg New York
ISBN-13   978-3-642-14639-8 Springer Berlin Heidelberg New York

springer.com

© Springer-Verlag Berlin Heidelberg 2010
Printed in Germany

Typesetting: Camera-ready by author, data conversion by Scientific Publishing Services, Chennai, India
Printed on acid-free paper   06/3180

ICEC 2009
Nitesh Chawla
Nathalie Japkowicz
Zhi-Hua Zhou
QIMIE 2009
Stéphane Lallich
Philippe Lenca
AIBDM 2009
Junbin Gao
Paul Kwan
Josiah Poon
Simon Poon
OSDM 2009
Peter Christen
GrahamWilliams
PAKDD Thai Track 2009
Thanaruk Theeramunkong
Cholwich Nattee
Boonserm Kijsirikul
Chotirat Ann Ratanamahatana
PAKDD 2009Workshop Arrangement Chair
Manabu Okumura
Bernhard Pfahringe
PAKDD 2009 Data Mining Competition
Paulo J. L. Adeodato

Preface

Five high-quality workshops were held at the 13th Pacific-Asia Conference on Knowledge Discovery and Data Mining (PAKDD 2009) in Bangkok, Thailand during April 27-30, 2009. There were 17, 6, 9, 4 and 5 accepted papers to be presented at the Pacific Asia Workshop on Intelligence and Security Informatics (PAISI 2009), the workshop on Advances and Issues in Biomedical Data Mining (AIBDM 2009), the workshop on Data Mining with Imbalanced Classes and Error Cost (ICEC 2009), the workshop on Open Source in Data Mining (OSDM 2009), and the workshop on Quality Issues, Measures of Interestingness and Evaluation of Data Mining Models (QIMIE 2009). One competition, PAKDD 2009 Data Mining Competition, and one local workshop, Thai Track Session, were arranged. From these workshops (except PAISI which published its works in separate LNCS proceedings), we selected two or three best papers for this LNCS publication. PAKDD is a major international conference in the areas of data mining (DM) and knowledge discovery in database (KDD). It provides an international forum for researchers and industry practitioners to share their new ideas, original research results and practical development experiences from all KDD-related areas including data mining, data warehousing, machine learning, databases, statistics, knowledge acquisition and automatic scientific discovery, data visualization, causal induction and knowledge-based systems.

In general, we wish to thank our General Workshop Co-chairs, Manabu Okumura and Bernhard Pfahringe, for selecting and coordinating the great workshops. We would like to thank Junbin Gao (Charles Sturt University), Paul Kwan (University of New England, Australia), Josiah Poon (University of Sydney), and Simon Poon (University of Sydney), for their arrangement of AIBDM 2009. We thank our ICEC 2009 Program Committee, Nitesh Chawla (University of Notre Dame), Nathalie Japkowicz (University of Ottawa), and Zhi-Hua Zhou (Nanjing University). We appreciate the OSDM Workshop Committee, Peter Christen (The Australian National University), and Graham Williams (Togaware, Australia) for his good arrangement work at the OSDM 2009. We also thank to the QIMIE Committee, Philippe Lenca and Stephane Lallich, for their arrangement of QIMIE 2009. We thank the PAKDD 2009 Data Mining Competition Committee, led by Paulo J. L. Adeodato. Three excellent papers were selected from the PAKDD 2009 Thai Track Session to be published in this LNCS volume.

The PAKDD 2009 workshops would not have been successful without the support of Program Committee members, reviewers, workshop organizers, invited speakers, organizing staff, and supporting organizations. Last but not least, specially thanks to our Organizing Committee members, including KIND laboratory at SIIT, Thammasat University, Nattapong Tongtep, Juniar Ganis, Peerasak Intarapaiboon, Jakkrit TeCho, Nichnan Kittiphattanabawon, Piya Limcharoen, for the publication of the PAKDD Workshops proceedings in the series of *Lecture*

Notes in Computer Science, and to Wirat Chinnan and Swit Phuvipadawat for their support of the PAKDD 2009 conference and workshop website. While the arrangement of the PAKDD 2009 conference and workshop involved so many people, we would like to extend an additional thank you to the contributors who helped with the PAKDD 2009 conference and workshop but their names may not be listed.

We greatly appreciate the support from various institutions. The conference was organized by the Sirindhorn International Institute of Technology (SIIT), Thammasat University (TU) and co-organized by the Department of Computer Engineering, Faculty of Engineering, Chulalongkorn University (CU), and Asian Institute of Technology (AIT). It was sponsored by the National Electronics and Computer Technology Center (NECTEC, Thailand), the Thailand Convention and Exhibition Bureau (TCEB), and the Air Force Office of Scientific Research/Asian Office of Aerospace Research and Development (AFOSR/AOARD). Finally, we wish to thank all authors and all conference participants for their contribution and support.

March 2010

Thanaruk Theeramunkong
Cholwich Nattee
Josiah Poon
Nitesh Chawla
Philippe Lenca
Peter Christen
Graham Williams
Paulo J. L. Adeodato

Organization

General Workshop Chair

Manabu Okumura   Tokyo Institute of Technology, Japan
Bernhard Pfahringe   University of Waikato, New Zealand

Data Mining When Classes Are Imbalanced and Errors Have Costs (ICEC 2009)

Organizing Committee

Nitesh Chawla   University of Notre Dame, USA
Nathalie Japkowicz   University of Ottawa, Canada
Zhi-Hua Zhou   Nanjing University, China

Publication and Web Chair

David Cieslak   University of Notre Dame, USA

Program Committee

Gustavo Batista   University of Sao Paulo, Brazil
Sanjay Chawla   University of Sydney, Australia
David Cieslak   University of Notre Dame, USA
Chris Drummond   National Research Council, Canada
Seyda Ertekin   Penn State University, USA
George Forman   HP Labs, USA
Robert Holte   University of Alberta, Canada
W. Philip Kegelmeyer   Sandia National Labs, USA
Taghi M. Khoshgoftaar   Florida Atlantic University, USA
Alek Kolcz   Microsoft Research, USA
Miroslav Kubat   University of Miami, USA
Charles Ling   University of Waterloo, Canada
Xu-Ying Liu   Nanjing University, China
Dragos Margineantu   Boeing Phantom Works, USA
Stan Matwin   University of Ottawa, Canada
Yuchun Tang   McAfee, Inc.
Gary Weiss   Fordham University, USA

Quality Issues, Measures of Interestingness and Evaluation of Data Mining Models (QIMIE 2009)

Organizing Committee

Stéphane Lallich ERIC	Universit Lyon 2, France
Philippe Lenca Lab-STICC	TELECOM Bretagne, France

Program Committee

Hidenao Abe	Japan
Jérôme Azé	France
José L., Balcázar	Spain
Bruno Crémilleux	France
Sven Crone	UK
Jean Diatta	La Réunion
Thanh-Nghi Do	Vietnam
Salvatore Gréco	Italy
Fabrice Guillet	France
Michael Hahsler	USA
Howard Hamilton	Canada
Martin Holena	Czech Republic
Stéphane Lallich	France
Ludovic Lebart	France
Philippe Lenca	France
Ming Li	China
Patrick Meyer	France
Annie Morin	France
David Olson	USA
Jan Rauch	Czech Republic
Gilbert Ritschard	Switzerland
Wanchai Rivepiboon	Thailand
Gilbert Saporta	France
Roman Slowinski	Poland
Robert Stahlbock	Germany
Athasit Surarerks	Thailand
Shusaku Tsumoto	Japan
Kitsana Waiyamai	Thailand
Dianhui Wang	Australia
Louis Wehenkel	Belgium
Gary Weiss	USA
Takahira Yamaguchi	Japan
Min-Ling Zhang	China
Djamel Zighed	France

Advances and Issues in Biomedical Data Mining (AIBDM 2009)

Organizing Committee

Junbin Gao Charles	Sturt University, Australia
Paul Kwan	University of New England, Australia
Josiah Poon	University of Sydney, Australia
Simon Poon	University of Sydney, Australia

Program Committee

Rafael Calvo	University of Sydney, Australia
Peter Christen	Australian National University, Australia
Dao-qing Dai	Sun Yat-Sen (Zhongshan) University, China
Kin Fun Li	University of Victoria, Canada
Steve Gunn	University of Southampton, UK
David Hansen	CSIRO, Australia
Irena Koprinska	University of Sydney, Australia
Ong Kok Leong	Deakin University, Australia
Wenyuan Li	University of Southern California, USA
Christine O'Keefe	CSIRO, Australia
Georg Peters	University of Applied Sciences Muenchen, Germany
Daming Shi	Nanyang Technological University, Singapore
Chunhua Weng	Columbia University, USA
Jun Zhang Huazhong	University of Science and Technology, China

The First Open Source in Data Mining Workshop (OSDM 2009)

Organizing Committee

Peter Christen	The Australian National University, Australia
Graham Williams	Togaware, Australia

Program Committee

Rohan Baxter	The Australian Taxation Office, Australia
Michael Berthold	University of Konstanz, Germany
Christian Borgelt	European Center for Soft Computing, Spain
Janez Demsar	University of Ljubljana, Slovenia
Eibe Frank	University of Waikato, New Zealand
Mark Hall	Pentaho, New Zealand
Joshua Huang	The University of Hong Kong, Hong Kong
Bernhard Pfahringer	University of Waikato, New Zealand
Blaz Zupan	University of Ljubljana, Slovenia
Yunming Ye	Harbin Institute of Technology, China

PAKDD 2009 Data Mining Competition

Organizing Committee

Paulo J. L. Adeodato (Chair)	NeuroTech Ltd. and Federal University of Pernambuco, Brazil
Adrian L. Arnaud (Vice-Chair)	NeuroTech Ltd., Brazil
Rosalvo F. Oliveira Neto	NeuroTech Ltd., Brazil
Fábio C. Pereira	NeuroTech Ltd., Brazil
Osmar V. Cunha Júnior	NeuroTech Ltd., Brazil
David J. Ribeiro	NeuroTech Ltd., Brazil
Icamaan B. V. Silva	NeuroTech Ltd. and Federal University of Pernambuco, Brazil
Domingos S. M. P. Monteiro	NeuroTech Ltd., Brazil
Pasakorn Tangchanachaianan	Chulalongkorn University, Thailand

Scientific Committee

Paulo J. L. Adeodato (Chair)	NeuroTech Ltd. and Federal University of Pernambuco, Brazil
Adrian L. Arnaud (Vice-Chair)	NeuroTech Ltd., Brazil
Germano C. Vasconcelos	NeuroTech Ltd. and Federal University of Pernambuco, Brazil
Rodrigo C. L. V. Cunha	NeuroTech Ltd., Brazil

PAKDD 2009 Thai Track Session

Program Committee

Thanaruk Theeramunkong	SIIT Thammasat University
Boonserm Kijsirikul	Chulalongkorn University
Cholwich Nattee	SIIT Thammasat University
Choochart Haruechaiyasak	NECTEC
Chotirat Ratanamatan	Chulalongkorn University
Chularat Tanprasert	NECTEC
Chutima Pisarn	Prince of Songkla University
Krisana Chinnasarn	Burapha University
Kitsana Waiyamai	Kasetsart University
Sanparith Marukatat	NECTEC
Thepchai Supnithi	NECTEC
Vilas Wuwongset	Asian Institute of Technology
Nittaya Kerdprasop	Mahasarakham University
Kittisak Kerdprasop	Mahasarakham University
Verayuth Lertnattee	Silpakorn University

Table of Contents

The iZi Project: Easy Prototyping of Interesting Pattern Mining
Algorithms.. 1
    Frédéric Flouvat, Fabien De Marchi, and Jean-Marc Petit

CODE: A Data Complexity Framework for Imbalanced Datasets....... 16
    Cheng G. Weng and Josiah Poon

An Empirical Study of Applying Ensembles of Heterogeneous Classifiers
on Imperfect Data 28
    Kuo-Wei Hsu and Jaideep Srivastava

Undersampling Approach for Imbalanced Training Sets and Induction
from Multi-label Text-Categorization Domains...................... 40
    Sareewan Dendamrongvit and Miroslav Kubat

Adaptive Methods for Classification in Arbitrarily Imbalanced and
Drifting Data Streams 53
    Ryan N. Lichtenwalter and Nitesh V. Chawla

Two Measures of Objective Novelty in Association Rule Mining........ 76
    José L. Balcázar

PAKDD Data Mining Competition 2009: New Ways of Using Known
Methods 99
    Chaim Linhart, Guy Harari, Sharon Abramovich, and Altina Buchris

Feature Selection for Brain-Computer Interfaces 106
    Irena Koprinska

Mining Protein Interactions from Text Using Convolution Kernels...... 118
    Ramanathan Narayanan, Sanchit Misra, Simon Lin, and
    Alok Choudhary

Missing Phrase Recovering by Combining Forward and Backward
Phrase Translation Tables....................................... 130
    Peerachet Porkaew and Thepchai Supnithi

Automatic Extraction of Thai-English Term Translations and
Synonyms from Medical Web Using Iterative Candidate Generation
with Association Measures 141
    Kobkrit Viriyayudhakorn, Thanaruk Theeramunkong,
    Cholwich Nattee, Thepchai Supnithi, and Manabu Okumura

Accurate Subsequence Matching on Data Stream under Time Warping
Distance 156
   *Vit Niennattrakul, Dechawut Wanichsan, and
   Chotirat Ann Ratanamahatana*

Author Index 169

The iZi Project: Easy Prototyping of Interesting Pattern Mining Algorithms

Frédéric Flouvat[1], Fabien De Marchi[2], and Jean-Marc Petit[2]

[1] University of New Caledonia, PPME, F-98851, Noumea, New Caledonia
frederic.flouvat@univ-nc.nc
[2] Université de Lyon, CNRS
Université Lyon 1, LIRIS, UMR5205, F-69621, France
fabien.demarchi@liris.cnrs.fr
[3] Université de Lyon, CNRS
INSA-Lyon, LIRIS, UMR5205, F-69621, France
jean-marc.petit@insa-lyon.fr

Abstract. In the last decade, many data mining tools have been developed. They address most of the classical data mining problems such as classification, clustering or pattern mining. However, providing classical solutions for classical problems is not always sufficient.

This is especially true for pattern mining problems known to be "representable as set", an important class of problems which have many applications such as in data mining, in databases, in artificial intelligence, or in software engineering. A common idea is to say that solutions devised so far for classical pattern mining problems, such as frequent itemset mining, should be useful to answer these tasks. Unfortunately, it seems rather optimistic to envision the application of most of publicly available tools even for closely related problems.

In this context, the main contribution of this paper is to propose a modular and efficient tool in which users can easily adapt and control several pattern mining algorithms. From a theoretical point of view, this work takes advantage of the common theoretical background of pattern mining problems isomorphic to boolean lattices. This tool, a C++ library called *iZi*, has been devised and applied to several problems such as itemset mining, constraint mining in relational databases, and query rewriting in data integration systems. According to our first results, the programs obtained using the library have very interesting performance characteristics regarding simplicity of their development. The library is open source and freely available on the Web.

1 Introduction

In the last decade, many data mining tools have been developed [1]: standalone algorithm implementations [2,3], packages [4], libraries [5], complete softwares with GUI [6,7] or inductive databases prototypes [8,9] . They address most of the classical data mining problems such as classification, clustering or pattern mining.

T. Theeramunkong et al. (Eds.): PAKDD Workshops 2009, LNAI 5669, pp. 1–15, 2010.
© Springer-Verlag Berlin Heidelberg 2010

However, providing classical solutions for classical problems is not always sufficient. For example, frequent itemset mining (FIM) is a classical data mining problems with applications in many domains. Many algorithms and tools have been proposed to solve this problem. Moreover, several works, such as [10], shown that FIM algorithms can be used as a building block for other, more sophisticated pattern mining problems. This is especially true for pattern mining problems known to be "representable as set" [10], an important class of problem which have many applications such as in data mining (e.g. frequent itemset mining and variants [11,12]), in databases (e.g. functional or inclusion dependency inference [13,14]), in artificial intelligence (e.g. learning monotone boolean function [15]), or in software engineering (e.g. software bug mining [16]).

In this setting, a common idea is to say that algorithms devised so far should be useful to answer these tasks. Unfortunately, it seems rather optimistic to envision the application of most of publicly available tools for frequent itemset mining, even for closely related problems. For example, frequent essential itemset mining [17] (as well as other conjunctions of anti-monotone properties) is very closely related to FIM. Actually, only the predicate test is different. In the same way, mining keys in a relational database is a pattern mining problem where, from a theoretical point of view, FIM strategies could be used. However, in both cases, users can hardly adapt existing tools to their specific requirements, and have to re-implement the whole algorithms.

Paper contribution. In this context, the main contribution of this paper is to propose a modular and efficient tool in which users can easily adapt and control several pattern mining algorithms. From a theoretical point of view, this work takes advantage of the common theoretical background of pattern mining problems isomorphic to boolean lattices. This tool, a C++ library called *iZi*, has been devised and applied to several problems such as itemset mining, constraint mining in relational databases and query rewriting in data integration systems. According to our first results, the programs obtained using the library have very interesting performance performance characteristics regarding simplicity of their development. The library is open source and freely available on the Web.

Paper organization. Section 2 discusses the value of our proposition w.r.t. existing related works. Section 3 introduces the *iZi* library. This section presents the underlined theoretical framework, points out how state of the art solutions can be exploited in our generic context, and describes the architecture of the *iZi* library. A demonstration scenario is presented in Section 4. Experimentations are described in Section 5. The last section concludes and gives some perspectives of this work.

2   Related Works

One may notice that algorithm implementations for pattern mining problems are "home-made" programs, see for example implementations available in FIMI workshops [2,3].

Packages, libraries, software, inductive databases prototypes have also been proposed, for instance Illimine [4], DMTL [5], Weka [6], ConQuest [8] and   [9].

Except DMTL, they provide classical algorithms for several data mining tasks (classification, clustering, itemset mining...). However, their algorithms are very specific and could not be used to solve equivalent or closely related problems. For example, even if most of these tools implement an itemset mining algorithm, none of them can deal with other interesting pattern discovery problems. Moreover, their source codes are not always available.

DMTL (Data Mining Template Library) is a C++ library composed of algorithms and data structures optimized for frequent pattern mining. Different types of frequent patterns (sets, sequences, trees and graphs) using generic algorithms implementations are available. Actually, DMTL supports any types of patterns representable as graphs. Moreover, the data is decoupled of the algorithms, and can be stored in memory, files, Gigabase databases (an embedded object relational database), and PSTL [18] components (a library of persistent containers). This library currently implements an exploration strategy: a depth-first approach (*eclat*-like [19]). Moreover, some support for breadth-first strategies is also provided. These algorithms could be used to mine all the frequent patterns of a given database.

To our knowledge, only the DMTL library has objectives close to *iZi*. Even if objectives are relatively similar w.r.t. code reusability and genericity, the motivations are quite different: while DMTL focuses on patterns genericity w.r.t. the frequency criteria only, *iZi* focuses on a different class of patterns but on a wider class of predicates. Moreover, *iZi* is based on a well established theoretical framework, whereas DMTL does not rely on such a theoretical background. However, DMTL encompasses problems that cannot be integrated into *iZi*, for instance frequent sequences or graphs mining since such problems are not isomorphic to a boolean lattice. The iZi library is complementary to DMTL since it offers the following new functionalities:

1. any monotone predicate can be integrated in *iZi*, while DMTL "only" offers support for the "frequent" predicate;
2. the structure of the patterns does not matter for *iZi*, while the patterns studied by DMTL must be representable as graphs (e.g. inclusion dependencies cannot be represented in DMTL);
3. while DMTL only gives all frequent patterns, *iZi* is able to supply different borders of "interesting" patterns (positive and negative borders). These borders are the solutions of many pattern mining problems. Moreover, end-users often do not care about all the patterns and prefer a smaller representation of the solution.

3   A Generic and Modular Solution for Patterns Discovery

3.1   A Generic Theoretical Framework

The theoretical framework of [10] formalizes *enumeration problems under constraints*, i.e. of the form "enumerate all the patterns that satisfy a condition".

When the condition must be verified in a data set, the word "enumerate" is commonly replaced by "extract". Frequently, the problem specification requires that patterns must be maximal or minimal w.r.t. some natural order over patterns.

Consequently, common characteristics of these problems are: 1) the predicate defining the interestingness criteria is monotone (or anti-monotone) with respect to a partial order $\preceq$ over patterns, 2) there exists a bijective function f from the set of patterns to a boolean lattice and its inverse f^{-1} is computable, and 3) the partial order among patterns is preserved, i.e. $X \preceq Y \Leftrightarrow f(X) \subseteq f(Y)$.

3.2   Algorithms

The classical way to solve pattern mining problems is to develop ad-hoc solutions from scratch, with specialized data structures and optimization techniques. If such a solution leads to efficient programs in general, it requires a huge amount of work to obtain a sound and operational program. Moreover, if problem specifications slightly change over time, a consequent effort should be made to identify what parts of the program should be updated.

One of our goal is to factorize some algorithmic solutions which can be common to any pattern mining problem representable as sets.

Currently, many algorithms from the multitude that has been proposed for the FIM problem could be generalized and implemented in a modular way, from well knowns *Apriori* algorithm [20] or depth-first approaches, to more sophisticated dualization-based algorithms (*Dualize and Advance* [21] or *ABS* [14,22]).

However, some algorithms don't fit in this framework because they are not based on a clear distinction between the exploration strategy and the problem. For example, FP-growth like algorithms [23] cannot be used into this framework since their strategy is based a data structure specially devised for FIM. In the same way, condensed representations based algorithms like LCM [24] cannot be applied to any pattern mining problem representable as sets.

The need to have multiple strategies in a pattern mining tool is twofold. First, note that the type of solution discovered by each algorithm is specific. For example, the *Apriori* algorithm discover (without any overhead) the theory and the two borders, whereas dualization-based algorithms "only" discovers the two borders. Since depending on the studied problem, we might be interested in either the theory, or the positive border, or the negative border, it is necessary to have multiple strategies to enable the discovery of the required solution. Secondly, as shown by the FIMI workshops, the algorithms performance depend on dataset/problem characteristics. For example, the *Apriori* algorithm is more appropriate when the theory is composed of relatively small elements, i.e. solutions are small patterns. Consequently, several algorithms must be integrated into a pattern mining tool to have the best performances according to problem properties.

3.3   The iZi Library

Based on the theoretical framework introduced in Section 3.1, we propose a C++ library, called *iZi*, for this family of problems. The basic idea is to offer a toolkit

providing several efficient, robust, generic and modular algorithm implementations. The development of this library takes advantage of our past experience to
solve particular pattern mining problems such as frequent itemsets mining, functional dependencies mining, inclusion dependencies mining and query rewriting.

Architecture. Figure 1 represents the architecture and the "workflow" of our
library: The *algorithm* is initialized (*initialization* component) with patterns
corresponding to singletons in the set representation, using the data (*data access*
component). Then, during the execution of the algorithm, the *predicate* is used
to test each pattern against the data. Before testing an element, the algorithm
use the *set transformation* component to transform each set generated into the
corresponding pattern.

This architecture is directly derived from the studied framework and has the
main advantage of decoupling algorithms, patterns and data. Only the *predicate*,
set transformation and *initialization* components are specifics to a given problem. Consequently, to solve a new problem, users may have to implement or reuse
with light modifications some of these components.

The *algorithm* component represents generic algorithm implementations provided with the library and used to solve pattern mining problems. As shown
in Figure 1, algorithms are **decoupled from the problems** and are a **black
box for users**. Each algorithm can be used directly to solve any problem fitting
in the framework without modifications. This leads to the rapid construction
of robust programs without having to deal with low level details. Currently,
the library offers a levelwise algorithm [10], a dualization-based algorithm, and
two other variants of these algorithms. These variants globally have the same
strategy but explore the search space in a different way (top-down exploration
instead of bottom-up) which is more appropriate for some predicates. Finally,
depth-first strategies are also currently being integrated.

Another important aspect of our library is that data access is totally decoupled
of all other components (see Figure 1). Currently, data access in most of the
other implementations is tightly coupled with algorithm implementations and

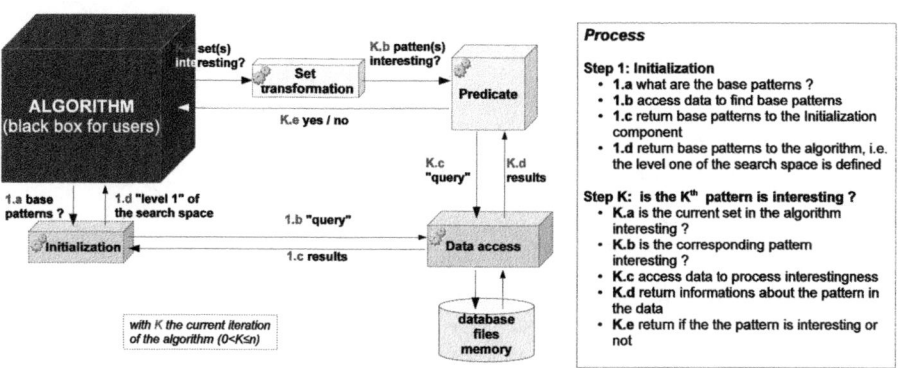

Fig. 1. *iZi* "workflow"

predicates. Consequently, algorithms and "problem" components can be used with different data formats without modifications.

Figure 2 presents how the library works for the IND (INclusion Dependency) mining problem. We suppose in this example that the algorithm used is the levelwise strategy.

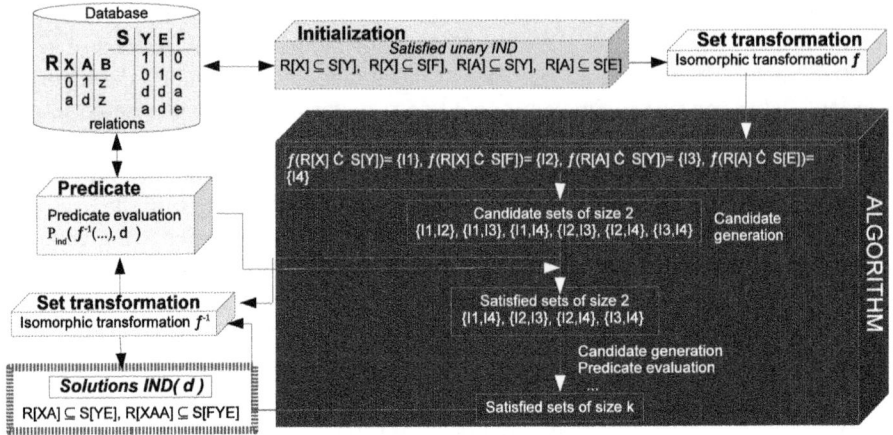

Fig. 2. IND mining example

Data structures. Since internally each algorithm only manipulates sets, we use a data structure based on prefix-tree (or trie) specially devoted to this purpose [25]. For example, Figure 3 represents the prefix-tree data structure associated to the set $\{\{A, C\}, \{A, D, F\}, \{A, D, G\}, \{A, E\}, \{D, E\}, \{E, F, G\}, \{E, G\}\}$.

They have not only a power of compression by factorizing common prefix in a set collection, but are also very efficient for candidate generation. Moreover, prefix-trees are well adapted for inclusion and intersection tests, which are basic operations when considering sets. Of course, as for algorithms, one can imagine

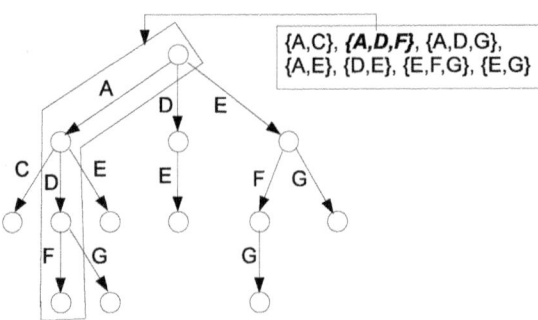

Fig. 3. Example of trie data structure

to extend our library with alternative structures for sets, like bitmaps. The use of indexes is also an important issue but not considered yet.

Note that template trie container and iterator are provided with the library. Actually, two trie implementations are available with the library: one optimized for data compression and one optimized for data search. Their implementation have been mapped on the implementations of the standard C++ STL (Standard Template Library) containers. This class also contains an implementation of an incremental algorithm, based on trie data structures, for the minimal transversals computation of an hypergraph.

Implementation issues. Figure 4 presents, from an implementation point of view, a UML model of the library. In particular, this model specifies how patterns and sets interact with the other components: patterns are used in problem specific components and sets are used internally by the algorithms. This model also points out the possibility to do predicate composition which is the case in many applications (e.g. itemset mining using conjunction of monotone constraints). For data access, this model distinguishes two cases: input data and output data. Input data is used by the predicate to test patterns and is totally independent of the algorithms. Output data is used by the algorithm to output the solutions (theory and/or positive border and/or negative border).

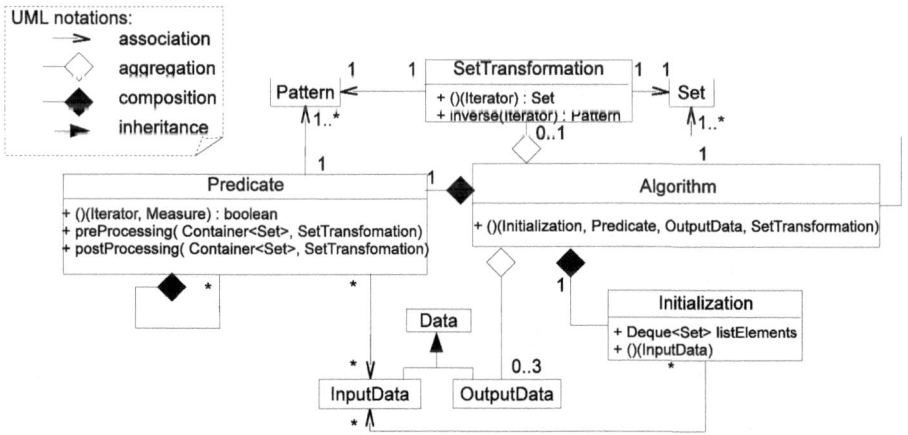

Fig. 4. iZi UML model

Moreover, thanks to this model and to the object-oriented paradigm, users can also implement algorithm and predicate variants/refinements, i.e. use inheritance to define new algorithms or predicate based on existing ones. Figure 5 presents an example of algorithm and predicate variants/refinements already implemented. In this figure, the *frequent* class represents the frequent itemset mining predicate, and the *frequent essential* class represents the predicate for a condensed representation of frequent itemsets. In the same way, the *Dualization* class represents the dualization based algorithm provided in *iZi* and the *ReverseDualization* class represents a variant of this algorithm changing the exploration strategy.

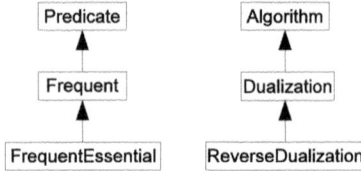

Fig. 5. Example of algorithm and predicate variants/refinements

In our context, another interesting property is method overloading which can be used to optimize some predicates sub-methods w.r.t. specific data structures. For example, the support counting method is crucial for frequent itemset mining algorithms. Using method overloading, it is for example possible to have a generic support counting method and one optimized for trie data structures. Thanks to this property, it is possible to have a good trade-off between components genericity and algorithms performances.

Finally, to solve a new problem, users only have few implementations constraints. For example, their *predicate* and *set transformation* classes have to be functors (i.e. function objects) with the same signature for the *operator()* method, and *output data* classes must only have a *push_back()* method. To facilitate these developments, abstract base classes are provided with the library as well as sample components.

4   Demonstration Scenario

From our past experience in the development of pattern mining algorithms, we note that the adaptation of existing implementations is extremely difficult. In some cases, it implies the redevelopment of most of the implementation and could take more time than developing a new program from scratch.

As shown in Table 1, many problems have been implemented in our library along with several data types and data sources components. For itemset mining, the format considered is the FIMI file format which has been defined by the FIMI workshops [2,3] to store transactional databases in a text file. This data format is widely used for this family of problems. For constraint mining in relational databases, components have also been developed to access data in files

Table 1. Problems and data sources experimented with *iZi*

Data type	Problem	File (format)	DBMS
tabular	inclusion dependencies [14]	CVS FDEP [26]	MySQL
tabular	keys [10]	CVS FDEP [26]	MySQL
binary	frequent itemsets [11]	FIMI [2,3]	
binary	frequent essential itemsets [17]	FIMI [2,3]	
set	sub-problems of query rewriting [27]	specific	

(CSV format of *Excel* and *FDEP* format defined by [26]) and in the *MySQL* DBMS. For query rewriting in integration systems, we have studied two combinatorial sub-problems (i.e. two different predicates). The data access and output components processed specific file formats.

As indication, the use of our library to implement a program for the key mining problem in a relational database has been done in less than one working day. Based on these components, the following scenario will show the simplicity of solving a new problem using *iZi*.

Let us suppose that a user wants to solve a new pattern mining problem using *iZi*: for example inclusion dependencies mining. **First, the user has to check some theoretical aspects:**

1. **Is the problem an "enumeration problem under constraint"?**
   An *inclusion dependency* (IND) is an expression of the form $R[X] \subseteq S[Y]$, where R and S are relation schemas of a same database schema D. Such a constraint ensures that, for any relations r and s over R and S, any X-value into r is a Y-value into s . If Y is a key in S, then X is a *foreign key* in R. Inclusion dependency discovery is a way to discover foreign keys and other more general semantic constraints. It can be stated as follows:

   IND mining problem (referred to as IND): Let d be a database over a schema D, extract (maximal) inclusion dependencies satisfied in d. Let $IND(d) = \{R[X] \subseteq S[Y] \mid R, S \in D, R[X] \subseteq S[Y]$ is satisfied in $d\}$.

2. **What are the patterns, the partial order and the predicate? Is the predicate (anti-)monotone?**
   a. The pattern language $\mathcal{L}_{ind}$ is composed of all the IND expressions that can be expressed into a database schema.
   b. The predicate $P_{ind}(R[X] \subseteq S[Y], d)$ is true, if $\pi_X(r) \subseteq \pi_Y(s)$ (with π the projection operator of the relational algebra).
   c. From a well known inference rule for INDs [28], if an IND is satisfied, then any IND obtained by applying the same projection on the left and right-hand sides is satisfied. As an example, if $R[ABC] \subseteq S[EFG]$ is satisfied, then the following INDs (not exhaustive) are satisfied: $R[A] \subseteq S[E], R[B] \subseteq S[F], R[C] \subseteq S[G], R[BC] \subseteq S[FG], R[CB] \subseteq S[GF]$... Consequently, the partial order is defined by projections over INDs.

   Considering the partial order defined by projections over INDs, the predicate $P_{ind}(R[X] \subseteq S[Y], d)$ is anti-monotone (see [14] for the proof).

   The IND mining problem can be reformulated as follows [14]:

   $$IND(d) = \mathcal{B}d^+(Th(\mathcal{L}_{ind}, d, P_{ind}))$$

3. **What is the function f that guarantees the isomorphism with a boolean lattice ?** (see [14] for more details on this point)
   The search space of IND is not a boolean lattice at all. As an example, consider the two INDs $R[X] \subseteq S[Y]$ and $R[X'] \subseteq T[Z]$. They do not have an

upper bound (i.e. a common specialization), such as $R[XX'] \subseteq S[YZ]$ for $R[X] \subseteq S[Y]$ and $R[X'] \subseteq S[Z]$, since they don't consider the same relations. To solve this, we have to consider the subproblems $IND(r, s)$ for each pair of relations {r,s} in d. However, the search spaces of these subproblems are still not boolean lattices. For example $R[A] \subseteq S[E]$ and $R[B] \subseteq S[F]$ have two possible least upper bound, which are $R[AB] \subseteq S[EF]$ and $R[BA] \subseteq S[FE]$. In order to fit each subproblem into a boolean lattice context, we define the function f which transforms any IND into the set of all unary INDs (i.e. INDs between single attributes) obtained by projection. Thus, $f(R[AB] \subseteq S[EF]) = \{R[A] \subseteq S[E]; R[B] \subseteq S[F]\}$. Now, the desirable property is that f must be a bijection between IND search space and the powerset of all unary INDs. However:

- f is not a one-to-one function, since $f(R[AB] \subseteq S[EF]) = f(R[BA] \subseteq S[FE])$. The solution is to restrict the IND search space to INDs with a sorted left-hand side. Thanks to the "permutation inference rule", this restriction leads to no loss of knowledge [28].
- f is not surjective, since e.g. $f^{-1}(\{R[A] \subseteq S[E]; R[B] \subseteq T[G]\})$ cannot be defined. To cope with this problem, one needs to mine INDs from pairs of relations one by one. Moreover, duplicate attributes must be allowed in IND definition as it is done in [29].

With the above restrictions, one can easily verify that f is an isomorphism between IND search space and the powerset of unary INDs.

The search space C of INDs over (R, S) is defined by: $C(R, S) = \{R[< A_1...A_n >] \subseteq S[< B_1...B_n >] \mid \forall 1 \leq i < j \leq n, (A_i < A_j) \vee (A_i = A_j \wedge B_i < B_j)\}$ where $n = min(|R|, |S|)$.

Let I_1 be the set of unary INDs over R. The function $f : C \longrightarrow \mathcal{P}(I_1)$ is defined by: $f(i) = \{j \in I_1 \mid j \preceq i\}$. The function $f : C \longrightarrow \mathcal{P}(I_1)$ is bijective and its inverse function f^{-1} is computable. Moreover, given i and j two IND expressions of C, $i \preceq j \Leftrightarrow f(i) \subseteq f(j)$.

Consequently, f is an isomorphism from $(C, \preceq)$ to $(\mathcal{P}(I_1), \subseteq)$, that is to say that the search space of INDs is representable as sets.

Let $\mathcal{L}_{ind} = C(R, S)$, the search space of $IND(r, s)$ is isomorphic to a boolean lattice, and the function f is $f : C \longrightarrow \mathcal{P}(I_1)$ (see [14] for the proof).

This example is a typical case: the problem becomes representable as sets by restricting the language to be used to define the search space (without any loss of knowledge thanks to patterns properties).

Secondly, the user has to develop (or adapt) several components:

4. the *data access* component. Suppose in this scenario that the data is stored in a MySQL database, and that a component for this data source is already implemented.
5. the *initialization* component, which will initialize unary INDs using databases schemas.

6. **the *set transformation* component**, which will transform an IND in a set of unary INDs (and inversely).
7. **the *predicate* component**, which will test if the IND in parameter is satisfied in the database (using the *data access* component).

Note that as shown by their source code, all these components are simple with few lines of code. Moreover, if some of them are already developed, the user can directly reuse them without modifications. See as an example Figure 6 for an implementation of the predicate component for IND mining in a MySQL database.

```
template< class DBMS>
class SatisfiedIND: public Predicate
{
 protected:
   //! Pointer on the dbms
   DBMS * mydbms;

 public:
   //! Constructor
   /*!
     \param inDbms pointer on the DBMS and the db studied
   */
   SatisfiedIND_DBMS( DBMS * inDbms )
   {
     mydbms = inDbms ;

     if( mydbms->get_relation(1) && mydbms->get_relation(2) )
     {
       // store a parameretrized query  to test inclusion
       // dependencies between the two input relations
       string query = "select count(*) from "
         + mydbms->get_relation(1)->name
         + " where ( var1 ) not in( SELECT distinct var2 FROM "
         + mydbms->get_relation(2)->name +" )" ;
       mydbms->store_query( (char *)(query.c_str()) );
     }
   }

   //! Operator that test if an inclusion dependency is satisfied
   //! or not in two relations
   /*!
     \param itCand iterator on the pattern to test
     \param mesCand measure associated wih the pattern and
processed in the predicate
   */
   template< class Iterator, class Measure >
   bool operator() ( Iterator  itCand, Measure & mesCand );
};
```

```
#include "SatisfiedIND.h"

//! Operator that test if an inclusion dependency is satisfied or
//! not in two relations
template< class DBMS>
template< class Iterator, class Measure >

bool SatisfiedIND<DBMS>::operator()(Iterator  itCand,
Measure & mesCand)
{
   //search the attributes in the left part of the IND
   string left= itCand->left[0] ;
   for( int i = 1; i < itCand->left.size(); i++)
           left+=","+itCand->left[i] ;

   //search the attributes in the right part of the IND
   string right= itCand->right[0] ;
   for( int i = 1; i < itCand->right.size(); i++)
           right+=","+itCand->right[i] ;

   left= "  "+left+" ";
   right=" "+right+" ";

   //replace the variables by the attributes of the IND
   mydbms->replace_in_query(" var1 ", (char *)(left.c_str()) );
   mydbms->replace_in_query(" var2 ", (char *)(right.c_str()) );

   //execute the query
   string nb_notin = mydbms->exec_query();

   //re initialize the variables for the next predicate test
   mydbms->replace_in_query( (char *)(left.c_str())," var1 ");
   mydbms->replace_in_query( (char *)(right.c_str())," var2 ");

   // test if  values of the first projection are in the second one
   if( nb_notin == "0" )   return true ;  // the IND is satisfied
   else  return false ;
}
```

Fig. 6. Example of IND predicate implementation

From this moment, the user can directly use any algorithm provided with *iZi* in his/her source codes, compile and execute the algorithm to find all satisfied INDs.

5   Experimentations

Our motivation here is to show that our generic library has good performance characteristics w.r.t. specialized and optimized implementations.

We present some experimental results for frequent itemset mining, since it is the original application domain of the algorithms we used and the only common problem with DMTL. Moreover, many resources (algorithms implementations, datasets, benchmarks...) are available on Internet [30] for frequent itemset mining. For other problems such as key mining, even if algorithms implementations are sometimes available, it is difficult to have access to the datasets. As an example, we plan to compare *iZi* with the proposal in [31] for key mining. Unfortunately, neither their implementation, nor their datasets have been made available in time.

Implementations for frequent itemset mining are very optimized, specialized, and consequently very competitive. The best performing ones are often the results of many years of research and development. In this context, our experimentations aims at proving that our generic algorithms implementations behave well compared to specialized ones. Moreover, we compare *iZi* to the DTML library, which is also optimized for frequent pattern mining.

Experiments have been done on some FIMI datasets [2,3] on a pentium 4.3GHz processor, with 1 Go of memory. The operating system was Ubuntu Linux 6.06 and we used gcc 4.0.3 for the compilation. We compared our *Apriori* generic implementation to two others devoted implementations: one by B. Goethals [32] and one by C. Borgelt [33]. The first one is a quite natural version, while the second one is, to our knowledge, the best existing *Apriori* implementation, developed in *C* and strongly optimized. Then, we compared "*iZi Apriori*" and "*iZi* dualization based algorithm" to the eclat implementation provided with DMTL.

In Table 2, three *Apriori* implementations are compared w.r.t. their execution times (in milliseconds) for datasets *Connect* (129 items and 67 557 transactions), *Pumsb* (2 113 items and 49 046 transactions) and *Pumsb** (2 088 items and 49 046 transactions). One can observe that our generic version has good performance with respect to other implementations. These results are very encouraging, in regards of the simplicity to obtain an operational program.

In Table 3, iZi and DMTL are compared w.r.t. their execution times (in milliseconds) for the same datasets. Even if DMTL is optimized and specialized for the frequent predicate, algorithm implementations of *iZi* have good performances w.r.t. eclat DMTL . The difference between the two libraries is mainly due to the algorithm used during the experimentations. This could be easily confirmed by looking at the performances of *Apriori*, *Eclat* and *dualization based* algorithms observed during FIMI benchmarks [30].

Table 2. Comparison of three *Apriori* implementations (in milliseconds)

	Apriori iZi	Apriori Goethals	Apriori Borgelt
Connect 90%	23 000	133 000	1 000
Pumsb 90%	18 000	14 000	1 000
Pumsb* 60%	2 000	4 000	1 000

Table 3. Comparison of iZi and DMTL implementations (in milliseconds)

	Apriori iZi	ABS iZi	eclat DMTL
Connect 90%	23 000	8 000	17 000
Pumsb 90%	18 000	18 000	8 000
Pumsb* 60%	2 000	2 000	5 000

6   Discussion and Perspectives

In this paper, we have considered a classical problem in data mining: the discovery of interesting patterns for problems known to be *representable as sets*, i.e. isomorphic to a boolean lattice. In addition to the interest of our library to solve new problems, *iZi* is also very interesting for algorithm benchmarking. Indeed, thanks to the modularity of *iZi*, it is possible to test several data representations (e.g. prefix tree or bitmap) or several predicates, with the same algorithm source code. Thus, it enables a fair comparison and test of new strategies. *iZi* has also been used for educational purpose. Using the library, students can better understand where the key issues are in pattern mining. For example, for frequent itemset mining, they often underestimate the importance of support counting in the algorithm performance. By allowing to easily change the strategy used for support counting, *iZi* enables to better understand how this affects algorithms performances.

To our knowledge, this is the first contribution trying to bridge the gap between fundamental studies in data mining around inductive databases [10,21,34] and practical aspects of pattern mining discovery. Our work concerns plenty of applications from different areas such as databases, data mining, or machine learning.

Many perspectives exist for this work. First, we may try to integrate the notion of *closure* which appears under different flavors in many problems. The basic research around concept lattices [35] could be a unifying framework. Secondly, we are interested in integrating the library as a plugin for a data mining software such as Weka [6]. Analysts could directly use the algorithms to solve already implemented problems or new problems by dynamically loading their own components. Finally, a natural perspective of this work is to develop a declarative version for such mining problems using query optimization techniques developed in databases [36].

References

1. Goethals, B., Nijssen, S., Zaki, M.J.: Open source data mining: workshop report. SIGKDD Explorations 7, 143–144 (2005)
2. Bayardo Jr., R.J., Zaki, M.J.(eds.): FIMI 2003, Proceedings of the IEEE ICDM Workshop on Frequent Itemset Mining Implementations, Melbourne, Florida, USA, November 19. CEUR Workshop Proceedings, vol. 90, CEUR-WS.org (2003)
3. Bayardo Jr., R.J., Goethals, B., Zaki, M.J. (eds.): FIMI 2004, Proceedings of the IEEE ICDM Workshop on Frequent Itemset Mining Implementations, Brighton, UK, November 1. CEUR Workshop Proceedings, vol. 126, CEUR-WS.org (2004)
4. Han, J.: Data Mining Group: IlliMine project. University of Illinois Urbana-Champaign Database and Information Systems Laboratory (2005), http://illimine.cs.uiuc.edu/
5. Hasan, M., Chaoji, V., Salem, S., Parimi, N., Zaki, M.: DMTL: A generic data mining template library. In: Workshop on Library-Centric Software Design (LCSD 2005), at OOPSLA 2005 conference, San Diego, California (2005)

6. Witten, I.H., Frank, E.: Data Mining: Practical machine learning tools and techniques, 2nd edn. Morgan Kaufmann, San Francisco (2005)
7. Mierswa, I., Wurst, M., Klinkenberg, R., Scholz, M., Euler, T.: Yale: rapid prototyping for complex data mining tasks. In: Eliassi-Rad, T., Ungar, L.H., Craven, M., Gunopulos, D. (eds.) KDD, pp. 935–940. ACM, New York (2006)
8. Bonchi, F., Giannotti, F., Lucchese, C., Orlando, S., Perego, R., Trasarti, R.: Conquest: a constraint-based querying system for exploratory pattern discovery. In: Liu, L., Reuter, A., Whang, K.Y., Zhang, J. (eds.) ICDE, p. 159. IEEE Computer Society, Los Alamitos (2006)
9. Blockeel, H., Calders, T., Fromont, É., Goethals, B., Prado, A., Robardet, C.: An inductive database prototype based on virtual mining views. In: Li, Y., Liu, B., Sarawagi, S. (eds.) KDD, pp. 1061–1064. ACM, New York (2008)
10. Mannila, H., Toivonen, H.: Levelwise search and borders of theories in knowledge discovery. Data Min. Knowl. Discov. 1, 241–258 (1997)
11. Agrawal, R., Imielinski, T., Swami, A.N.: Mining association rules between sets of items in large databases. In: Buneman, P., Jajodia, S. (eds.) SIGMOD Conference, pp. 207–216. ACM Press, New York (1993)
12. Mannila, H., Toivonen, H.: Multiple uses of frequent sets and condensed representations (extended abstract). In: KDD, pp. 189–194 (1996)
13. Koeller, A., Rundensteiner, E.A.: Heuristic strategies for inclusion dependency discovery. In: Meersman, R., Tari, Z. (eds.) OTM 2004, Part II. LNCS, vol. 3291, pp. 891–908. Springer, Heidelberg (2004)
14. De Marchi, F., Flouvat, F., Petit, J.M.: Adaptive strategies for mining the positive border of interesting patterns: Application to inclusion dependencies in databases. In: Boulicaut, J.-F., De Raedt, L., Mannila, H. (eds.) Constraint-Based Mining and Inductive Databases. LNCS (LNAI), vol. 3848, pp. 81–101. Springer, Heidelberg (2006)
15. Angluin, D.: Queries and concept learning. Machine Learning 2, 319–342 (1987)
16. Li, Z., Zhou, Y.: Pr-miner: automatically extracting implicit programming rules and detecting violations in large software code. In: Wermelinger, M., Gall, H. (eds.) ESEC/SIGSOFT FSE, pp. 306–315. ACM, New York (2005)
17. Casali, A., Cicchetti, R., Lakhal, L.: Essential patterns: A perfect cover of frequent patterns. In: Tjoa, A.M., Trujillo, J. (eds.) DaWaK 2005. LNCS, vol. 3589, pp. 428–437. Springer, Heidelberg (2005)
18. Gschwind, T.: Pstl-a c++ persistent standard template library. In: COOTS, pp. 147–158. USENIX (2001)
19. Zaki, M.J., Parthasarathy, S., Ogihara, M., Li, W.: New algorithms for fast discovery of association rules. In: KDD, pp. 283–286 (1997)
20. Agrawal, R., Srikant, R.: Fast algorithms for mining association rules in large databases. In: Bocca, J.B., Jarke, M., Zaniolo, C. (eds.) VLDB, pp. 487–499. Morgan Kaufmann, San Francisco (1994)
21. Gunopulos, D., Khardon, R., Mannila, H., Saluja, S., Toivonen, H., Sharm, R.S.: Discovering all most specific sentences. ACM Trans. Database Syst. 28, 140–174 (2003)
22. Flouvat, F., De Marchi, F., Petit, J.M.: ABS: Adaptive Borders Search of frequent itemsets. In: [3]
23. Han, J., Pei, J., Yin, Y.: Mining frequent patterns without candidate generation. In: Chen, W., Naughton, J.F., Bernstein, P.A. (eds.) SIGMOD Conference, pp. 1–12. ACM, New York (2000)
24. Uno, T., Asai, T., Uchida, Y., Arimura, H.: Lcm: An efficient algorithm for enumerating frequent closed item sets. In: [2]

25. Bodon, F.: Surprising results of trie-based fim algorithms. In: [3]
26. Flach, P.A., Savnik, I.: Database dependency discovery: A machine learning approach. AI Commun. 12, 139–160 (1999)
27. Jaudoin, H., Flouvat, F., Petit, J.M., Toumani, F.: Towards a scalable query rewriting algorithm in presence of value constraints. Journal on Data Semantics 12, 37–65 (2009)
28. Abiteboul, S., Hull, R., Vianu, V.: Foundations of Databases. Addison-Wesley, Reading (1995)
29. Mitchell, J.C.: The implication problem for functional and inclusion dependencies. Information and Control 56, 154–173 (1983)
30. Goethals, B.: Frequent itemset mining implementations repository, http://fimi.cs.helsinki.fi/
31. Sismanis, Y., Brown, P., Haas, P.J., Reinwald, B.: Gordian: Efficient and scalable discovery of composite keys. In: Dayal, U., Whang, K.Y., Lomet, D.B., Alonso, G., Lohman, G.M., Kersten, M.L., Cha, S.K., Kim, Y.K. (eds.) VLDB, pp. 691–702. ACM, New York (2006)
32. Goethals, B.: Apriori implementation. University of Antwerp, http://www.adrem.ua.ac.be/~goethals/
33. Borgelt, C.: Recursion pruning for the apriori algorithm. In: [3]
34. Boulicaut, J.F., Klemettinen, M., Mannila, H.: Modeling kdd processes within the inductive database framework. In: Mohania, M.K., Tjoa, A.M. (eds.) DaWaK 1999. LNCS, vol. 1676, pp. 293–302. Springer, Heidelberg (1999)
35. Ganter, B., Wille, R.: Formal Concept Analysis. Springer, Heidelberg (1999)
36. Chaudhuri, S.: Data mining and database systems: Where is the intersection? IEEE Data Eng. Bull. 21, 4–8 (1998)

CODE: A Data Complexity Framework for Imbalanced Datasets

Cheng G. Weng and Josiah Poon

School of Information Technologies, J12
University of Sydney, NSW, 2006, Australia
{cheng,josiah}@it.usyd.edu.au

Abstract. Imbalanced datasets occur in many domains, such as fraud detection, cancer detection and web; and in such domains, the class of interest often concerns the rare occurring events. Thus it is important to have a good performance on these classes while maintaining a reasonable overall accuracy. Although imbalanced datasets can be difficult to learn, but in the previous researches, the skewed class distribution has been suggested to not necessarily being the one that poses problems for learning. Therefore, when the learning of the rare class becomes problematic, it does not imply that the skewed class distribution is the cause to blame, but rather that the imbalanced distribution may just be a byproduct of some other hidden intrinsic difficulties.

This paper tries to shade some light on this issue of learning from imbalanced dataset. We propose to use data complexity models to profile datasets in order to make connections with imbalanced datasets; this can potentially lead to better learning approaches. We have extended from our previous work with an improved implementation of the CODE framework in order to tackle a more difficult learning challenge. Despite the increased difficulty, CODE still enables a reasonable performance on profiling the data complexity of imbalanced datasets.

Keywords: Imbalanced datasets, Data complexity.

1   Introduction

The imbalanced dataset problem is an important research area that has received a growing amount of attention over the past few years. Motivations for solving some of the imbalanced dataset problems are attached with great economical values. The economical incentives come from problems like credit card fraud detection, cancer detection, and marketing selections. These problems can easily cost billions of dollar every year across the globe. The problem of imbalanced datasets happens because the conventional machine learners aim to maximize the overall accuracy, which will lead to a bias towards the majority class under imbalanced situations. However, with imbalanced datasets, the interesting class is often the rare class. Therefore, it is better to have a good performance on the rare class while maintaining a reasonable overall accuracy.

T. Theeramunkong et al. (Eds.): PAKDD Workshops 2009, LNAI 5669, pp. 16–27, 2010.
© Springer-Verlag Berlin Heidelberg 2010

In the imbalanced dataset community, it has been suggested that the skewed class distribution may be a consequence of another disguised intrinsic problem. Yet, there is still no systematic way of explaining what underlying causes are affecting the learning performance of an imbalanced dataset. We think the answer could be found through data complexity analysis.

In this paper, we continue to explore the data complexity models under imbalanced situations. In our previous work [16], we have proposed an alternative data complexity framework that tries to describe data complexity based on the local information of the problem rather than only taking a broad, global analysis, which can often be misleading due to the effect of the majority class. We have updated our work, and the new materials in this paper include: a more refined explanation of CODE with a slight name change, an improved implementation of CODE, as well as results from more real world datasets.

The next section discusses the related works, it is then followed by a section introducing our CODE framework. We describe our experimental setup in section 4, and these results will then be presented with discussions in section 5. Lastly, we will conclude by proposing some possible future investigations.

2   Related Works and Motivations

There have been many attempts to resolve the imbalanced dataset problem. The biggest problem that we see is the lack of overall guidance as to which cause needs to be solved in any imbalanced dataset. Researchers have been working on different methods and carrying out experiments in different domains, and still, we have yet to agree on standard benchmark datasets also on a systematic approach for resolving class imbalance problem. Currently, the most common approach for resolving class imbalance is the re-sampling strategy. Other approaches vary from modifying existing algorithms, cost sensitive learning, one-class learning, or some combination of the above [4,2,13].

It is suggested that class imbalance does not always hinder classification performance[11,8]. The problem seems to be related to learning with a small target class size when it is in the presence of other factors, e.g. class overlapping [10]. The experiments conducted by [9] have also suggested that the degradation in performance is not directly caused by class imbalances, but rather, the small disjuncts.

In an attempt to address the issue of how to learn from an imbalanced dataset, we will utilize data complexity framework. Data complexity in our context is defined as the degree of difficulty to learn from a dataset. The data complexity measurements are derived for quantifying the data complexity. Our approach is mainly geometrically-based, meaning we are looking at the location of data points and how they are distributed across the feature space, as well as the relationship between the data points in terms of their geometrical position in the feature space, e.g. their distances measured by Euclidean distance.

As far as we know, the relevant work on measuring data complexity is done by [6]. They composed a set of data complexity measurements and they were able to show a notable difference in the data structure between a real world dataset and a random noise dataset. They observed the existence of a relationship between the overall learning performance and the data complexity measurements. They further suggested the possible use of data complexity measurements to guide the dynamic selection of classifiers for certain problems. Their measurements provided a global view of data complexity of which they looked at data point distribution across feature space, noise level, amount of sub-clusters, distance ratio of intra-class and inter-class examples, the non-linearity of the dataset, and fisher discriminate ratio. However, the limitation of their work is their framework's inability to reflect the performance of the rare class due to the dominating effect of the majority class. There is another work done by [14], in which they have also looked at the data complexity model by [6] and tried to analyze how learner's performance can be explained by looking at data complexity measurements. However, they only looked at two datasets and did not apply any correlation analysis to make the relationship more explicit. Therefore, we have further explored data complexity in [16] with the CODE framework and also extending the work done by [6]. We have both compared the data complexity framework and carried out experiments with real world datasets; we have found some promising results from this, and they indicate that data complexity is capable of capturing the level of difficulty presented in imbalanced datasets.

The data complexity approach is related to the meta-learning research [12] because data complexity measurements can also be seen as meta-information. However, we have a different focus in this paper. While meta-learning is about exploiting meta-knowledge about learning in order to enhance the learning performance; our focus is to find a connection between the imbalanced dataset problem and data complexity. Therefore, we focus more on finding an explanation behind the problem rather than solving the problem; which in our opinion, is the first step to take towards answering difficult questions in the imbalanced dataset community. Questions like, when can an imbalanced dataset cause problems for learning, when to employ the over-sampling or the under-sampling approach, and an even more fundamental question is how to precisely define the imbalanced dataset problem instead of setting a threshold on the class distribution or size.

3   CODE

We have developed CODE, a new data complexity framework, as an alternative to the existing data complexity measurements. Our objective is to improve the previous data complexity model in terms of the computational cost as well as the ability to describe the data complexity of an imbalanced dataset. The main difference between CODE and the previous complexity measurements is that CODE utilizes local information of a dataset to capture the data complexity. The local information that we are referring to is in the sense of a region. Regardless of the referenced space, when we talk about local information it refers to an enclosure of space and the information about the enclosed space is considered to be a local information. In contrast to local information is the global information,

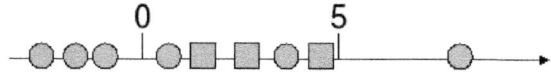

Fig. 1. Local v.s. global information

This simple example illustrates the difference between local and global information. The above 1D dataset contains 66.7% circle objects and 33.3% square objects, this is a *global information*; whereas the enclosed region between 0 and 5 contains 40% circle objects and 60% square objects, this is a *local information* about the referenced region.

which is the information collected from the whole dataset. Figure 1 is a simple illustration of the difference between the two different levels of information. Local information requires an enclosed region to generate information that is local to that referenced region. However, if we extend the enclosed space to cover the whole dataset, then global information can actually be local information as well, but this would probably defeat the purpose of having a more localized data complexity measurements. The acronym, CODE, stands for 4 concepts: *Cluster, Overlap, Density* and *Error*. These four concepts is designed to work together as a single unit so they are meant to compliment one another. We will go through each concept and describe how they work.

The first concept is *Cluster*, which plays a big part in the CODE framework. In this first concept, the dataset is divided into smaller groups so that we can collect local information on an imbalanced dataset to give a more appropriate and somewhat more representative data complexity measurements. The use of local information has been shown to be more resilient to the dictating influence of the majority class [16]. By restricting the space to more manageable parts is an advantage because it can potentially separate the difficult clusters from the easy ones, so the learning process could potentially made more flexible and adaptive.

Overlap, as the name suggests, is a measurement for detecting how much the classes have overlapped with each other. Intuitively, high degrees of class overlapping makes the dataset more difficult to learn. In literature, the effect of class overlapping in imbalanced datasets have been investigated [3,10], and from their experiments, we see there is a direct correlation between the two. So more overlapping tend to make the imbalanced dataset harder to learn. Therefore, the rationale is to quantify the level of difficulty in separating different class examples.

Density looks at the distribution of data points in the feature space, e.g. dense and sparse areas in the space. The concept is inspired by [18], where SVM have been reported to produce misaligned decision boundary for an imbalanced dataset due to the uneven distribution of the two classes. This phenomena will lead to more false negatives because SVM will place the "optimal" hyperplane much too close to the rare class examples. Therefore, the Density concept aims to help better understand the dynamics between the different distribution of data points in the dataset. By doing so, it can also be effective at capturing the intra v.s. inter class imbalance phenomenon that was discussed in the literature[7].

Error is a concept that looks at the level of noise in a dataset, which is meant to identify any irregularities, outliers, or potential errors in a dataset. This measurement tries to quantify how much noise is present in a dataset because it is generally agreed that the quality of a dataset has a direct impact on the upper bound of the learning performance. It has also been discussed in the literature that the noise is a contributing factor in the imbalanced datasets learning process [13]. Another way of thinking about this concept is to think about the coherence of the data, so if a group of data points has high coherence it would imply that there is a sense of harmony in the group because data points fits well together, which would lead to low noise level in the dataset.

CODE is designed to be a general conceptual framework, so there is no one fixed implementation and each concept in CODE can be implemented into different sets of practical data complexity measurements. In the next section we will describe an implementation of CODE.

3.1   CODE Updated

In our initial implementation of CODE, the first step clustering was done with k-mean, but in the updated version of CODE we uses EM instead. We have made the change because EM will automatically determine the number of clusters by running cross validation, and even though it requires more computation, but it eliminates the need to preset the number of clusters.

After clustering, we measure different attributes for their ability to separate different class examples. There are two types of attribute in our consideration: numeric and nominal attributes. For any dataset, we would like to measure the goodness of its best attribute for separating the two class. However, due to the difference of numeric and nominal attributes, we will perform different measurement for each type of attributes. In the end, base on the value computed, we take the best attribute, one for each attribute type, for separating the classes among all the attributes as the resulting measurement. The changes that we have incorporated in the updated version is to make adjustments to the formulas in order to make sure that the value is more consistent with the wording of "overlapping", which means the higher the value, the higher the overlapping.

For the numeric attributes, we use numeric overlapping formula *(numOverlap)*, which is the inverse of fisher discriminant ratio [6]. The formula for fisher discriminant ratio is shown in equation 1, where μ_1, μ_2, σ_1^2 and σ_2^2 are means and variances of the two classes:

$$fisher = \frac{(\mu_1 - \mu_2)^2}{\sigma_1^2 + \sigma_2^2} \tag{1}$$

and the reason for using the inverse of *fisher* is for consistency with the meaning of overlapping, which is the opposite of discriminant. Therefore, the numeric overlapping has the following formula:

$$numOverlap = \frac{1}{fisher} = \frac{\sigma_1^2 + \sigma_2^2}{(\mu_1 - \mu_2)^2} \tag{2}$$

The formula for nominal attribute was inspired by the value difference metric [17] and we have formulated it as in equation 3:

$$nomRatio_a = \sum_{v=0}^{n} \frac{minN_{a,v,c}}{maxN_{a,v,c}} \times \frac{N_{a,v}}{total} \tag{3}$$

$$nomOverlap = min([nomRatio_0, nomRatio_1, ... nomRatio_m]) \tag{4}$$

where $N_{a,v}$ is the number of instances in the training data that has value v for the attribute a. We can further split $N_{a,v}$ into different classes, writing as $N_{a,v,c}$, this means the number of class c instances with the value v for the attribute a. Since we are currently only deal with two-class problems, so $minN_{a,v,c}$ denotes the value for the class with a smaller $N_{a,v}$ and $maxN_{a,v,c}$ is the other class, which has a larger $N_{a,v}$. The reason for using the notation $minN_{a,v,c}$ and $maxN_{a,v,c}$ is because the class may not be the same for different values of attributes, so we cannot use a fixed value for c. After we compute the ratio of $minN_{a,v,c}$ and $maxN_{a,v,c}$, it is then weighted base on the presence of the attribute value in the dataset, i.e. if many instances has the same attribute value, the value gets a larger weight. *total* is the total number of instances in the dataset. We repeat this calculation for all values of attribute a and take the sum as $nomRatio_a$ of attribute a. Note, regardless of the class, the smaller count for an attribute value is treated as the numerator, because we are only interested in the ratio. We assign the $min(nomRatio_{0..a})$ as the nominal overlapping value. The range of $nomOverlap$ is from 0 to 1, smaller value means less class overlapping and the attribute will have more discriminative power

In addition to $numOverlap$ and $nomOverlap$, we have also added a new geometric-based overlapping measurement ($geoOverlap$), which is based on the idea of fisher discriminant ratio, but instead of performing calculations on the attribute space, we use the geometrical space instead. The calculation is as follows: the first step is to calculate the center for each class. It is the same process as in k-mean clustering, but we calculate the mean for each class rather than for each clusters. For each class we add up every example in the class and take the average, which can be expressed more precisely as:

$$m_c = \frac{1}{|S_c|} \sum_{x_i \in S_c} x_i \tag{5}$$

where m_c is the class center and S_c is the set of examples in class c.

After we have the centers, the second step is to calculate the standard deviation, in geometric terms, for each class. Here, we use Euclidean distance for the difference between two examples:

$$\sigma_c = \sqrt{\frac{1}{|S_c|} \sum_{x_i \in S_c} |x_i - m_c|^2} \tag{6}$$

The final step is to apply the same $numOverlap$ formula (eq. 2) to get the $geoOverlap$ of the two classes.

For density, we have kept the measure of class distribution for each clusters and the new addition is a geometric-based measure, which is the geometric standard deviation that we have calculated in the equation 6. This measurement is a measure of density because standard deviation measures the variability or dispersion of a population. If we have a small standard deviation, it means the data points are closer and more dense, whereas large standard deviation means that the data points are sparse and far away from each other.

Error concept in our context is referring to examples that are incoherent, that does not fit well with certain model. These errors can also be called the noise of the dataset. In concrete terms, a noisy example may be something that is classified incorrectly base on its nearest neighbour. In our previous implementation we use 1-nearest-neighbour to determine the level of noise, we record the accuracy of classifying examples within each cluster as the measure of noise level. In the updated implementation, we use Naive Bayes (NB) instead, because we have realised that this process is effectively the same as a leave-one-out cross-validation procedure. Therefore one can actually have the freedom to choose whichever model they believe would be a good model for the given dataset. We have chosen NB because it can run much faster than 1-nearest-neighbour.

The implementation described above still has the same shortcoming as the initial implementation, which is that the generated CODE measurements are not a uniform set of measurements. For different datasets, we may have different number of clusters, which will produce a variable number of CODE values. This is a problem if we want to perform supervised learning to find correlation between data complexity and learning performance. An ideal solution for this problem will be to generate an uniform set of features and still able to reflect most, if not all, of the local information that CODE model is trying to convey. In our current implementation, we have used the mean-variance approach, which is essentially taking the mean and variance of each measurements in CODE from each cluster. Although this averaging approach does not fully reflect the information that original CODE contains, but it does a reasonable job in our experiments. However, we will address this shortcoming in the future updates of CODE.

4   Experiments

Dataset

In our experiments, we have used 56 datasets from the UCI data repository [1]. Since our framework only work with binary problems and there is only a limited number of binary class problems available, so we have transformed multiclass problems with the 1-vs-others approach. Twenty-seven datasets were collected from UCI data repository: anneal, audiology, autos, balance-scale, breast-cancer, Wisconsin-breast-cancer, colic.ORIG, credit-a, credit-g, diabetes, glass, heart-Cleveland, heart-stalog, hepatitis, ionosphere, iris, kr-vs-kp, lymphography, segment, sonar, soybean, splice, vote, vehicle, vowel, waveform and zoo. After we transformed the multi-class problems, we end up with a total of 56 datasets.

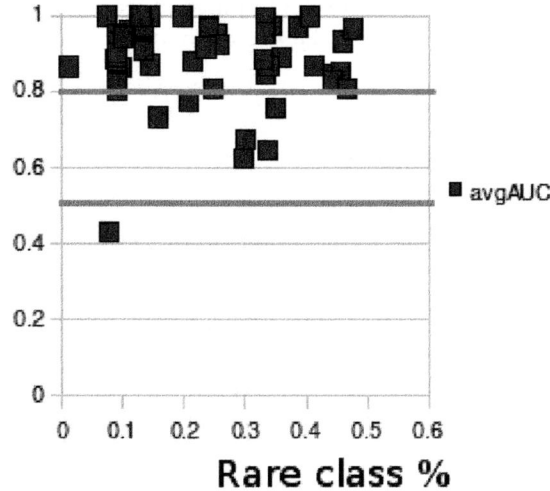

Fig. 2. Datasets: rare class distribution v.s. averaged AUC

This figure shows a plot of all the datasets with their corresponding rare class distribution (x-axis) and their averaged AUC score (y-axis). 0.5 is the baseline value for AUC and the best value is 1 or 0 (because one can inverse the decisions). 0.8 is a threshold that we use to separate "easy" datasets and "hard" datasets. Above 0.8 (or below 0.2) will be consider as easy, whereas between 0.2 to 0.8 is "hard".

Due to space restriction, we only show the rare class distribution of the dataset in figure 2, which also shows the averaged AUC performance across different learners.

Setup

In order to ensure the results are not too biased, we used 4 different learners, 2 linear and 2 non-linear learners. In the two linear learners, we have both discriminative and generative models, namely Decision Tree and Naive Bayes. As for the non-linear learners, we have the classic k-Nearest Neighbour and a kernel-based method, Support Vector Machine. First, we apply the learners on all the datasets and measure the learning performance on the smaller class with area under ROC curve (AUC). AUC is chosen because it has been found to be a useful measure for imbalanced dataset problems [5,15]. All experiments are done with 10 fold cross-validation and repeated 10 times to reduce the variance of the results. In the end, each dataset will have 4 AUC values from each learner, these AUC values are averaged to produce an averaged AUC score. We use this score as a general indicate of whether a dataset is easy or difficult to learn. The averaged AUC is shown in figure 2 with their respective rare class distribution. After we have computed all the AUC values, we combine them with the corresponding data complexity values and apply linear regression to portrait the relationship between the AUC and the data complexity of each dataset. 5 different linear regression models were built to correlate 4 different learner's AUC values, as well as the averaged AUC score, to the data complexity. The results are discussed in the next section.

5   Results and Discussion

The resulting correlations of data complexity and the learning performance were not as good as our previous experiment, however, it was expected. We has achieved 0.74, 0.59, 0.52, 0.6 and 0.67 correlation values respectively for 3-nearest-neighbour, decision tree, naive bayes, SVM and the average AUC. The correlation value is between 1 to -1, 1 means there is a strong positive correlation and -1 means strong inverse correlation, so values close to 0 would be a bad result because there is no obvious linear correlation found. Our results shows moderate correlations, which is still an encouraging results considering this is a more difficult task, and there is probably not enough data sample to achieve above 0.9 correlations.

In this paper, we have used 56 domains from the UCI data repository, which is 8 times more than before. It is more difficult because each domain can have a different data characteristics and we suspect that as the data diversity getting larger, so does the number of dataset sample required for building a good correlation model. This relationship is illustrated in figure 3, which shows a positive

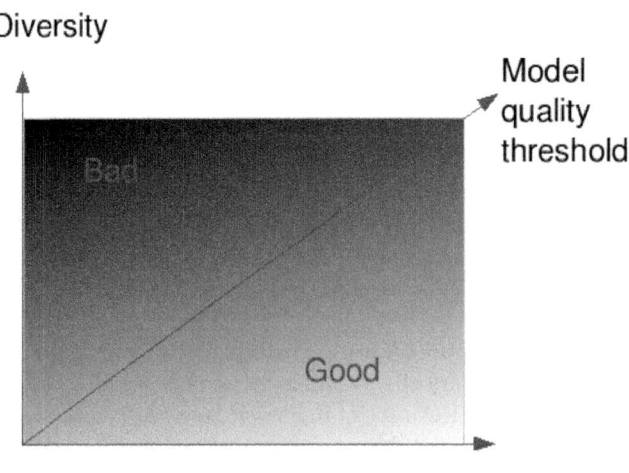

Fig. 3. Data complexity model quality

The quality of the model can be estimated from a direct relationship between the diversity of datasets and the number of datasets used. The gradient of colour shows the best model lies at bottom right-hand corner and the model quality gradually decreases as you move towards the top left-hand corner.

Table 1. balance-scale: "balanced"

cluster	numOlap	densityC0%	error0%	error1%	~
1	1	0.09	1	0	~
2	5.08	0.08	1	0	~
3	0.82	0.05	1	0	~

EM has found 3 clusters and we only have space to show some of the CODE values for this dataset. *cluster* is the cluster assignment, *numOlap* is the numeric overlapping, *densityC0%* is the rare class distribution within the cluster, *error0%* is the percentage of mis-classified rare class examples in the cluster and *error1%* is the mis-classification rate of the common class examples.

correlation between the diversity and the size of the dataset samples that we use to build the data complexity model. A similar analogy would be the relationship between dimensionality and the dataset size at a fixed level of data complexity; it would be more difficult to learn a good model from a high dimensional dataset if the training dataset is not have large enough. In theory, one can draw a threshold line to specify the required number of datasets in order to achieve a reasonable model quality, this line would be the theoretical lower bound. Finding this lower bound is an interesting but challenging research topic and we will not cover in this paper.

From figure 2, we can see an outlier which is the most difficult domain to learn. This dataset is the from the *balance-scale* dataset and the rare class label is called "balanced", which has a low 0.08 class distribution. So, why is this imbalanced dataset difficult to learn? This is where data complexity will help, we know it has a very skewed class distribution, but when we look closer at the CODE value, it has revealed some of the possible difficulties. With limited space, we only show some of its CODE values in table 1 and limit our discussion to these values. Here is what we have noticed:

1. There seems to be three natural clusters.
2. Cluster 2 seems to be more difficult than the other two clusters because it has higher numeric overlapping.
3. The presence of the rare class in each cluster are all quite low.
4. The rare class examples seems to be scattered across the three clusters.
5. The rare class examples all seem like noises in each cluster, because the NB model has got all the rare class examples wrong.

By reading CODE values, we can learn more information about the dataset than just knowing that it is an imbalanced dataset and it is difficult to learn. Knowing more about the domain that you are working with could be a great help in the data mining process.

6   Conclusion and Future Work

In this paper, we have described an improved version of CODE and the main contribution was that we have addressed a shortcoming in our previous work by using a more diverse set of datasets to find the relationship between data complexity and the learning performance. We have shown that CODE is able to capture a reasonable amount of data complexity despite the increased difficulty. We also shown how one can use CODE values to inspect different data characteristics that might have caused learning to become difficult. Some of the issues, such as the relationship between the data diversity and the required dataset sample size is also discussed.

Some future directions for this work would be to use synthetic datasets to build up a data complexity model and see if we can use it to describe different type of imbalanced datasets. Another interesting future work would be to visualize the datasets and visually verify some of the finding that CODE is producing. There are many potential usage for data complexity and we hope more research will be done to make data complexity more explicit and accessible for different data mining researches.

References

1. Asuncion, A., Newman, D.: UCI machine learning repository. University of California, Irvine, School of Information (2007)
2. Batista, G.E., Monard, M.C., Bazzan, A.L.C.: Improving rule induction precision for automated annotation by balancing skewed data sets. LNCS, pp. 20–32. Springer, Heidelberg (2004)
3. Batista, G.E.A.P.A., Prati, R.C., Monard, M.C.: Balancing Strategies and Class Overlapping. In: Famili, A.F., Kok, J.N., Peña, J.M., Siebes, A., Feelders, A. (eds.) IDA 2005. LNCS, vol. 3646, pp. 24–35. Springer, Heidelberg (2005)
4. Chawla, N.V., Bowyer, K.W., Hall, L.O., Kegelmeyer, W.P.: SMOTE: synthetic minority over-sampling technique. Journal of Artificial Intelligence Research 16, 321–357 (2002)
5. Fawcett, T.: ROC graphs: Notes and practical considerations for researchers. Machine Learning 31 (2004)
6. Ho, T.K., Basu, M.: Complexity measures of supervised classification problems. IEEE Transactions on Pattern Analysis and Machine Intelligence 24, 289–300 (2002)
7. Japkowicz, N.: Concept-Learning in the Presence of Between-Class and Within-Class Imbalances. In: Stroulia, E., Matwin, S. (eds.) Canadian AI 2001. LNCS (LNAI), vol. 2056, pp. 67–77. Springer, Heidelberg (2001)
8. Japkowicz, N.: Class imbalances: are we focusing on the right issue. In: Workshop on Learning from Imbalanced Data Sets II (2003)
9. Jo, T., Japkowicz, N.: Class Imbalances versus Small Disjuncts. SIGKDD Explor. Newsl. 6, 40–49 (2004)
10. Prati, R.C., Batista, G., Monard, M.C.: Learning with class skews and small disjuncts. LNCS, pp. 296–306. Springer, Heidelberg (2004)
11. Provost, F.: Machine Learning from Imbalanced Data Sets 101. In: AAAI Workshop on Learning from Imbalanced Data Sets. AAAI Press, Menlo Park (2000)
12. Vilalta, R., Giraud-Carrier, C., Brazdil, P., Soares, C.: Using Meta-Learning to Support Data Mining. International Journal of Computer Science& Applications 1, 31–45 (2004)

13. Weiss, G.M.: Mining with Rarity: A Unifying framework. SIGKDD Explor. Newsl. 6, 7–19 (2004)
14. Weng, C., Poon, J.: A Data Complexity analysis on imbalanced Datasets and an alternative imbalance Recovering Strategy. In: IEEE/WIC/ACM International Conference on Web Intelligence (2006)
15. Weng, C.G., Poon, J.: A New Evaluation Measure for Imbalanced Datasets. In: Seventh Australasian Data Mining Conference, vol. 87, pp. 27–32 (2008)
16. Weng, C.G., Poon, J.: Data Complexity Analysis for Imbalanced Datasets. In: PAKDD Workshop Data Mining When Classes are imbalanced and Errors have Costs, ICEC 2009 (2009)
17. Randall Wilson, D., Martinez, T.R.: Improved Heterogeneous Distance Functions. Journal of Artificial Intelligence Research 6, 1–34 (1997)
18. Wu, G., Chang, E.Y.: KBA: kernel boundary alignment considering imbalanced data distribution. IEEE Transactions on knowledge and data engineering 17, 786–795 (2005)

An Empirical Study of Applying Ensembles of Heterogeneous Classifiers on Imperfect Data

Kuo-Wei Hsu and Jaideep Srivastava

University of Minnesota, Minneapolis, MN, USA
{kuowei,Srivastava}@cs.umn.edu

Abstract. Two factors that slow down the deployment of classification or supervised learning in real-world situations. One is the reality that data are not perfect in practice, while the other is the fact that every technique has its own limits. Although there have been techniques developed to resolve issues about imperfectness of real-world data, there is no single one that outperforms all others and each such technique focuses on some types of imperfectness. Furthermore, quite a few works apply ensembles of heterogeneous classifiers to such situations. In this paper, we report a work on progress that studies the impact of heterogeneity on ensemble, especially focusing on the following aspects: diversity and classification quality for imbalanced data. Our goal is to evaluate how introducing heterogeneity into ensemble influences its behavior and performance.

Keywords: bagging, AdaBoost, diversity, heterogeneity, imbalanced data.

1 Introduction

Classification is an important data mining task and has been systematically studied for decades. Nevertheless, two factors slow down its deployment in more real-world situations. One is that most real-world data sets are not perfect [8, 23], while the other is that every classification technique has its own limits. In practice, data usually come with missing values, noise, and imbalanced distributions between classes or within a class. For the second factor, ensemble techniques are usually employed to overcome limitations of an individual classification technique. An ensemble creates a committee that collects predictions from all member classifiers and combines them to form final predictions. The motive for ensemble is similar to the concept that "different models excel in different regions" [11]. Ensemble compensates the limits of an individual classification technique. In some sense, this point of view is close to one, adopted in [9], that suggests "starting globally, optimizing locally, and predicting globally".

It is well known that diversity among member classifiers contributes to the success of ensemble. Moreover, using heterogeneous classifiers to create a committee would increase diversity [14]. Here, heterogeneous classifiers are those based on different algorithms. In this paper, we report our work in progress that studies the impact of

T. Theeramunkong et al. (Eds.): PAKDD Workshops 2009, LNAI 5669, pp. 28–39, 2010.

heterogeneity on ensemble, such as bagging and boosting (e.g., AdaBoost). Because distinguished researchers have done investigations into using ensembles to improve classification performance for difficult and imperfect data sets [4, 5, 6, 7, 20 24], ensembles show us an opportunity to review problems from imperfectness of data.

The rest of this paper is organized as follows. In Section 2, we will demonstrate motivational examples. In Section 3, we will examine the procedure described in [14] first and then extend the procedure for further investigations. Next, we will report results for the extension of classical bagging and AdaBoost algorithms in Section 4. Finally conclusions will be given in Section 5.

2 Examples

Diversity is one of factors contributing to the success of ensemble techniques. Because boosting tends to generate more diverse classifier combinations [26] and heterogeneity enhances diversity [14], we use a variant of AdaBoost and employ heterogeneous classifiers. Following the basic procedure of AdaBoost, the variant alternatively creates classifiers with different algorithms.

The first example is for synthetic data sets with multiple decision boundaries. Here we run experiments on six synthetic data sets (Eq. 61, p.37, in [12]), as summarized in the right panel of Figure 1. We use the programming libraries provided by WEKA [25]. We construct three ensembles each of which is composed of ten member classifiers. They are based on decision tree (DT), artificial neural network (ANN), and DT+ANN (i.e., the combination of DT and ANN), respectively. We use C4.5 [22] and multilayer perceptron for DT and ANN, respectively. Results reported here are from five-fold cross-validation. Note here that, we are not comparing an ensemble of heterogeneous algorithms with a single algorithm but with an ensemble that is composed of homogeneous algorithms. As shown in Fig. 1, DT outperforms ANN in less complicated data sets while ANN outperforms DT in data sets that present mixtures of linear and nonlinear transformation functions. Surprisingly, accuracies given by an ensemble composed of DT and ANN are not necessarily in between accuracies from using only DT and accuracies from using only ANN. In fact, the ensemble (DT+ANN) outperforms others even in complicated data sets.

The second example is for thirteen benchmark data sets from the UCI repository [3, 13]. It is associated with Table 1. Here we apply the variant of AdaBoost with three algorithms, DT, NB (Naïve Bayes [15]), and kNN (k-nearest neighbor [1]), on these data sets. We consider four ensembles each of which is composed of heterogeneous algorithms (HE), while we compare results from them with those from three ensembles each of which consists of homogeneous algorithms (HO). For each benchmark data set, we randomly draw samples from it with replacement and accordingly produce two data sets each of which is as large as it. We use one as the training set and the other as the test set. Test accuracies averaged over thirteen data sets are reported in Table 1, where ensembles of heterogeneous algorithms achieve the better stability by giving relatively smaller standard deviations.

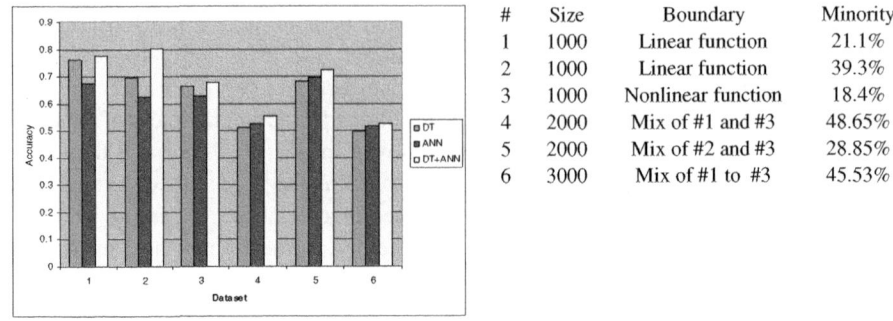

#	Size	Boundary	Minority
1	1000	Linear function	21.1%
2	1000	Linear function	39.3%
3	1000	Nonlinear function	18.4%
4	2000	Mix of #1 and #3	48.65%
5	2000	Mix of #2 and #3	28.85%
6	3000	Mix of #1 to #3	45.53%

Fig 1. Accuracy (left) for six synthetic data sets (right)

Table 1. Test accuracies (mean and standard deviations) on benchmark data sets

	Classifiers	10-iteration run	100-iteration run
HO	DT	0.764 ± 0.110	0.857 ± 0.137
	NB	0.758 ± 0.140	0.809 ± 0.094
	kNN	0.880 ± 0.079	0.924 ± 0.067
HE	DT+NB	0.893 ± 0.049	0.932 ± 0.080
	DT+kNN	0.892 ± 0.051	0.962 ± 0.023
	NB+kNN	0.954 ± 0.024	0.923 ± 0.091
	DT+NB+kNN	0.901 ± 0.037	0.971 ± 0.022

3 Diversity

Diversity measures are proposed to assist in the selection of member classifiers in an ensemble [17, 19]. However, because there is no widely accepted diversity measures [4, 36], this paper as well as the paper [14] consider several measures in the establishment of the quantitative determination of diversity among classifiers. Initially, ten popular diversity measures are Q-statistic (Q), correlation coefficient (ρ), double-fault measure (DF), interrater agreement (κ), measure of difficulty (θ), disagreement measure (DIS), entropy (E), Kohavi-Wolpert variance (KW), generalized diversity (GD), and coincident failure diversity (CFD). The paper [18] gives a succinct summary of them. Furthermore, we employ four other diversity measures to do further investigations: Weighted count of errors and correct results ($WCEC$) [2], Exponential Error Count (EEC) [2], Distinct failure diversity (DFD) [21], Measure of soft label outputs (M) [10].

   Furthermore, two synthetic data generators are employed to generate ten synthetic data sets. The number of samples in each data set is 10000 and the size of the feature set of each data set is 10. Neither missing values nor noise appear. In eight data sets percentages of minority class are from 42.09% to 49.91%, while those in the other two are 26.29% and 39.87%. We also use benchmark data sets. Considering three disparate algorithms, DT, NB, and kNN, we select any two of them and create a pair of heterogeneous classifiers. Therefore, for each data set, we set up six experimental sets in each of which we compare diversity of the combination of homogeneous classifiers with that of heterogeneous ones.

Table 2. Diversity in ten measures

Diversity values averaged over ten test data sets		Synthetic data sets		Benchmark data sets	
		Bagging setting		Bagging setting	
		HO	HE	HO	HE
Measures that prefer lower (absolute) values	Q	1	0.44	1	0.71
	ρ	1	0.16	1	0.29
	DF	0.22	0.08	0.12	0.05
	κ	1	0.44	1	0.71
	θ	0.13	0.08	0.15	0.12
Measures that prefer higher values	DIS	0	0.28	0	0.15
	E	0	0.28	0	0.15
	KW	0	0.07	0	0.04
	GD	0	0.7	0	0.66
	CFD	0	0.82	0	0.78

Table 3. Diversity in fourteen measures

Average diversity values		Synthetic data sets				Benchmark data sets			
		Bagging setting		Boosting setting		Bagging setting		Boosting setting	
		HO	HE	HO	HE	HO	HE	HO	HE
Measures that prefer lower (absolute) values	Q	1	0.58	-0.84	-0.74	1	0.78	-0.94	-0.93
	ρ	1	0.26	-0.26	-0.18	1	0.42	-0.24	-0.19
	DF	0.24	0.11	0.04	0.03	0.12	0.06	0	0
	κ	1	0.48	-0.08	0.04	1	0.75	0.04	0.15
	θ	0.12	0.07	0.06	0.06	0.16	0.13	0.08	0.08
	EEC	137	50	28	11	7	2	2	1
Measures that prefer higher values	DIS	0	0.26	0.54	0.48	0	0.12	0.48	0.43
	E	0	0.26	0.54	0.48	0	0.12	0.48	0.43
	KW	0	0.06	0.14	0.12	0	0.03	0.12	0.11
	GD	0	0.61	0.91	0.93	0	0.56	0.98	0.99
	CFD	0	0.74	0.95	0.96	0	0.69	0.99	0.99
	$WCEC$	-429	212	495	597	215	322	326	346
	DFD	0	0.01	0.48	0.42	0.02	0.06	0.75	0.79
	M	0	0.29	0.09	0.21	0	0.11	0.07	0.08

The procedure described in [14] is as follows: Given an input data set D and algorithms A_1 and A_2 (where $A_1 \neq A_2$), we sample D without replacement and generate two training sets. For synthetic data sets, we sample 10%; for benchmark data sets, we sample 50%. Next, we use A_1 and one training set to create (classifier) C_1. Then, we use A_1 and the other training set to create C_2. Next, we create C_3 by using the second training set and A_2. Then, we sample D with replacement and produce ten test sets. Finally, for each test set, we collect predictions and calculate the diversity between C_1 and C_2 (HO) and also that between C_1 and C_3 (HE). Results averaged over ten test sets are presented in Table 2, where each number represents a group of six experimental sets. Furthermore, we extend the above bagging-like setting to a boosting-like setting. More results are in Table 3.

We extend the above procedure and do exponential updates for weights of data samples based on the quality of predictions. Results are given in Table 3. Here we consider four more diversity measures, two more classification algorithms, and one more benchmark data set. Both bagging and boosting settings are with five algorithms: DT, NB, kNN, ANN, and SOM (Self-Organizing Map [16]). Data sets used here are the same as those used earlier, while one additional benchmark data set in used in this subsection such that here are fourteen benchmark data sets in total. Moreover, average diversity values reported in Table 3 are from fourteen diversity measures, among which ten are summarized in [18] and four are introduced earlier. In Table 3, diversity given by heterogeneous classifiers is still better than that given by homogeneous ones. However, such an advantage is not very dominative in boosting setting. This might be because that boosting would focus on difficult samples and its sampling procedure would act like an over-sampling procedure (for difficult samples) in later iterations.

4   Classification Quality

Based on the above discussion, we extend bagging and AdaBoost algorithms such that each randomly selects an algorithm (used as the base algorithm) in every iteration. This random selection strategy (for algorithm) is different from what we have seen in Section 2, where an alternative selection strategy (for algorithm) is employed. Moreover, we consider various metrics and report results from five-fold cross-validation. For balanced data sets, we simply present the error rates; for imbalanced data sets, we focus on minority and present results in false-positive rate (FPR), area under ROC (AUC), and F-measure.

1) Balanced and small data set (*sonar*: 208 samples, 60 features, and the minority is 47%). Curves in Fig. 2 show stable patterns in bagging but relatively unstable ones in boosting. This might be because of the small size of the data set.

2) Balanced and large data sets (*kr-vs-kp*: 3196 samples, 36 features, majority is 48%). In Fig. 3, the stabilities of curves from homogeneous and heterogeneous classifiers in boosting are much better than those shown in Fig. 2. This is because the data set is larger.

3) Imbalanced and small data (*hepatitis*: 155 samples, 19 features among 15 with missing values, the minority is 21% and labeled as "*a*"). For bagging, in terms of FPR in Fig.4, the best one is DT while the worst two are NB and kNN, and heterogeneous classifiers give curves in between; in terms of AUC in Fig. 5, the worst one is kNN while the best one is DT+NB. For boosting, since the size of data set plays an important role and *hepatitis* is a small data set, naturally it is not easy to see what classifier is the winner when the number of iterations is small. Results in F-measure for the minority are in Fig. 6.

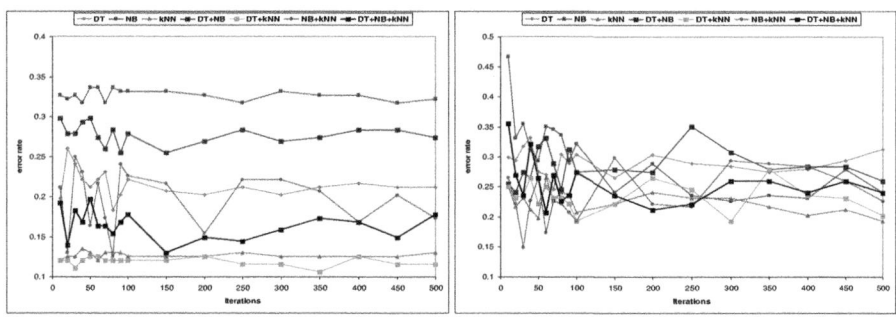

Fig. 2. Error rates vs. iterations for the data set *sonar* in bagging (left) and boosting (right)

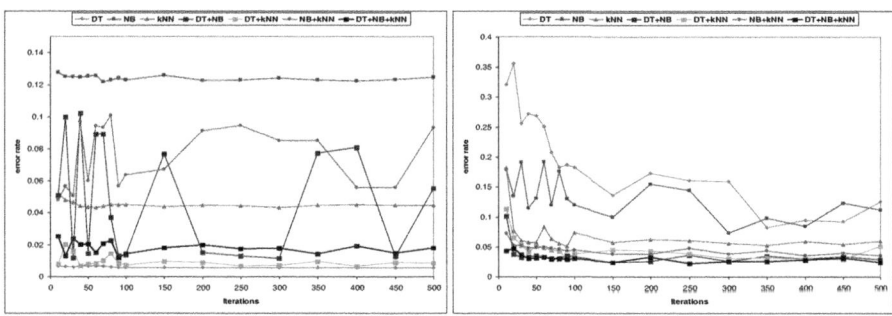

Fig. 3. Error rates vs. iterations for the data set *kr-vs-kp* in bagging (left) and boosting (right)

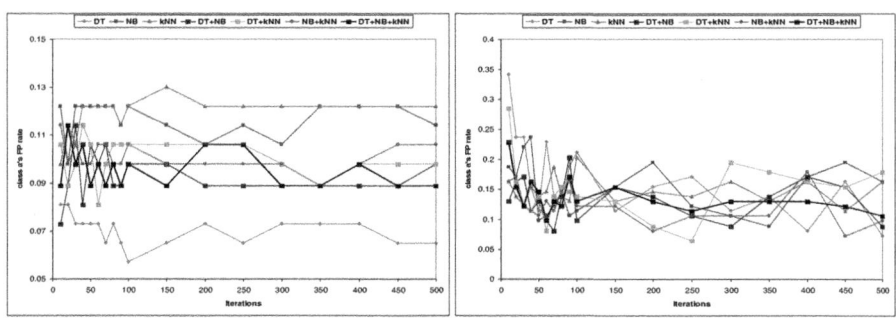

Fig. 4. FPR (minority) vs. iterations for the data set *hepatitis* in bagging (left) and boosting (right)

4) Imbalanced and large data (*sick*: 3772 samples, 29 features, among which 8 with missing values, the minority is 6% and labeled as "*b*"). For FPR shown in Fig. 7, the worst one is NB and the best one is DT+kNN in bagging, while worst ones are DT and NB in boosting. kNN is better than these two but not better than heterogeneous

classifiers. For AUC presented in Fig. 8, heterogeneous classifiers (except NB+kNN) provide good results in both bagging and boosting. Fig. 9 presents F-measure for the minority.

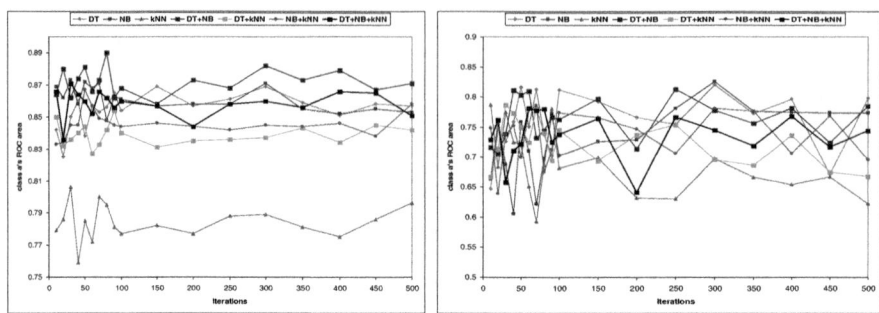

Fig. 5. AUC (minority) vs. iterations for the data *hepatitis* in bagging (left) and boosting (right)

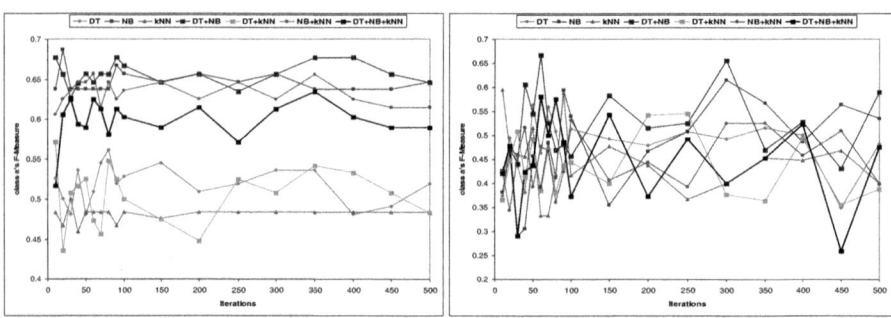

Fig. 6. F-measure (minority) vs. iterations for the data *hepatitis* in bagging (left) and boosting (right).

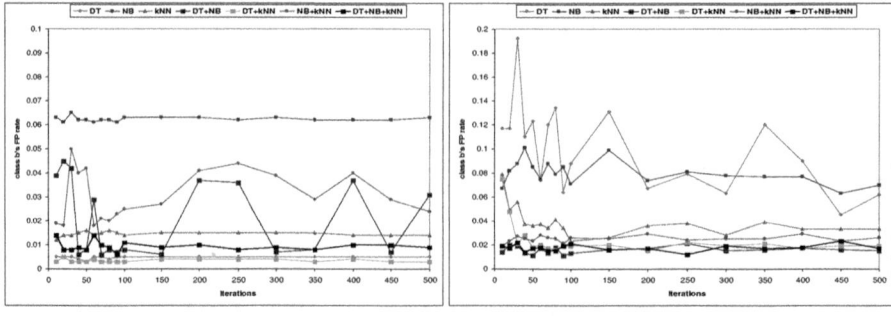

Fig. 7. FPR (minority) vs. iterations for the data set *sick* in bagging (left) and boosting (right).

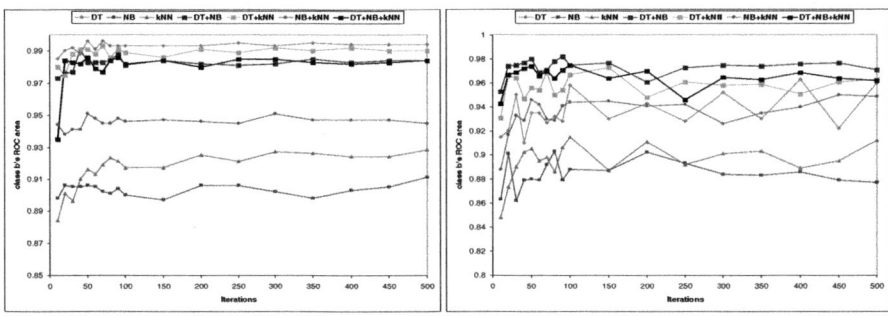

Fig. 8. AUC (minority) vs. iterations for the data *sick* in bagging (left) and boosting (right)

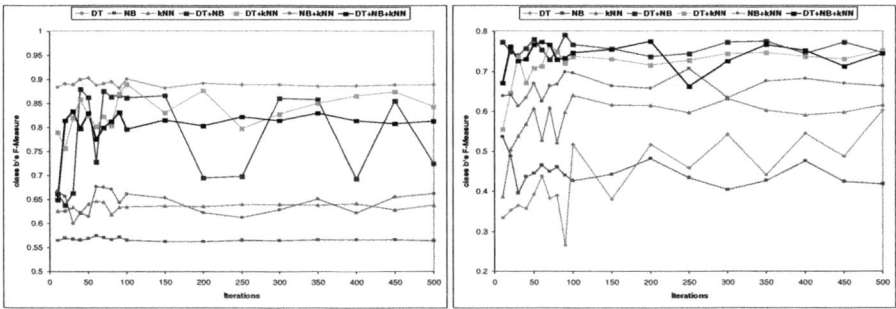

Fig. 9. F-measure (minority) vs. iterations for the data *sick* in bagging (left) and boosting (right)

Next, we consider data sets with noise: Results presented below are obtained from using the data set that contains artificially-added class noise in training and the original data set in test. They are with 100 iterations. For balanced data sets, we present error rates; for imbalanced ones, we present precision and recall for minority class.

1) Balanced and small data set (*sonar*). Fig. 10 shows the results. In bagging, NB is the worst while both DT (which is well known as an unstable algorithm) and kNN (which is a stable algorithm) show relatively stable curves (close to linear ones). Using DT and kNN in heterogeneous classifiers improves stability. For example, DT+NB is better than NB while NB+kNN is even better. Similar observations could be made in boosting. Moreover, DT+NB+kNN gives relatively stable curves in both bagging and boosting.

2) Balanced and large data set (*kr-vs-kp*). Fig. 11 shows the results. In bagging, NB+kNN is not only better than NB but also better than kNN. This suggests the usefulness of heterogeneity. In boosting, heterogeneous classifiers give relatively stable curves. Although DT is the worst, heterogeneous classifiers where it is used perform well.

3) Imbalanced and small data set (*hepatitis*). Results are shown in Fig. 12 and Fig. 13. For both precision and recall for minority class, it is expected that they decrease as the percentage of noise increases. Neither homogeneous nor homogeneous

classifiers perform well in boosting since the data set is small and AdaBoost is more sensitive to noise than bagging. If we look at 10% class label noise in bagging, NB+kNN performs well (but is not the best) even though NB is the worst in precision and kNN is the worst in recall.

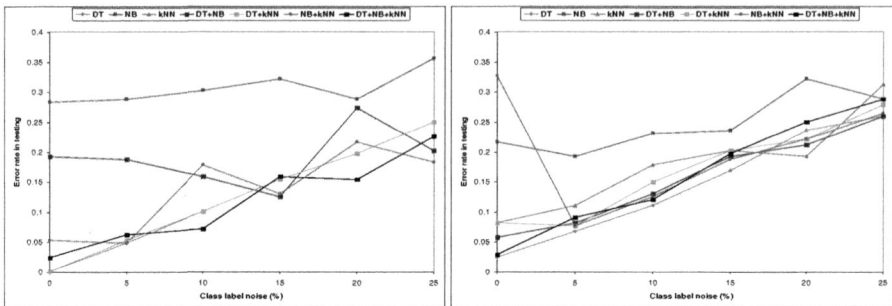

Fig. 10. Error rates vs. noise level for the data set *sonar* in bagging (left) and boosting (right).

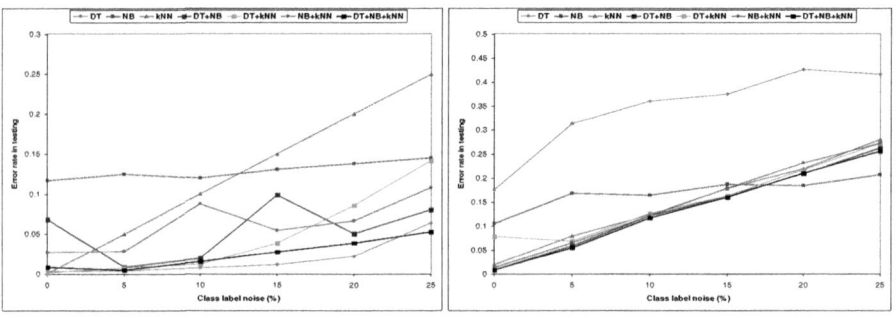

Fig. 11. Error rates vs. noise level for the data set *kr-vs-kp* in bagging (left) and boosting (right)

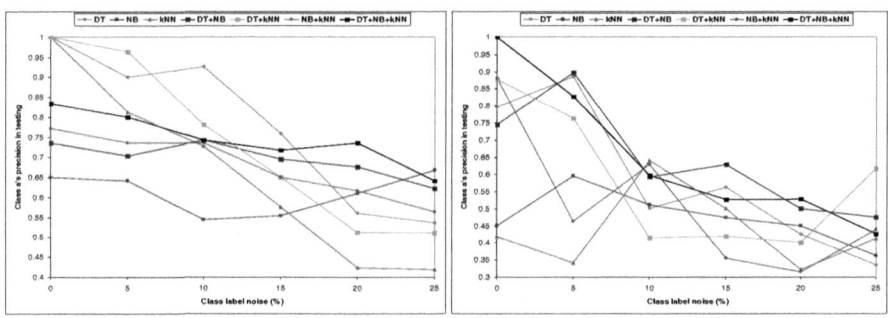

Fig. 12. Precision (minority) vs. noise level for *hepatitis* in bagging (left) and boosting (right)

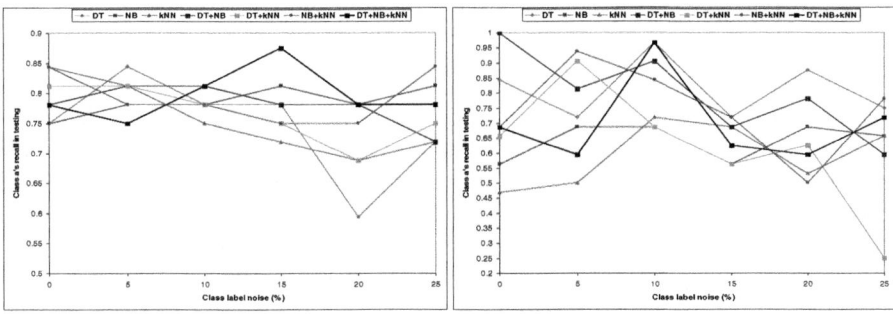

Fig. 13. Recall (minority) vs. noise level for *hepatitis* in bagging (left) and boosting (right)

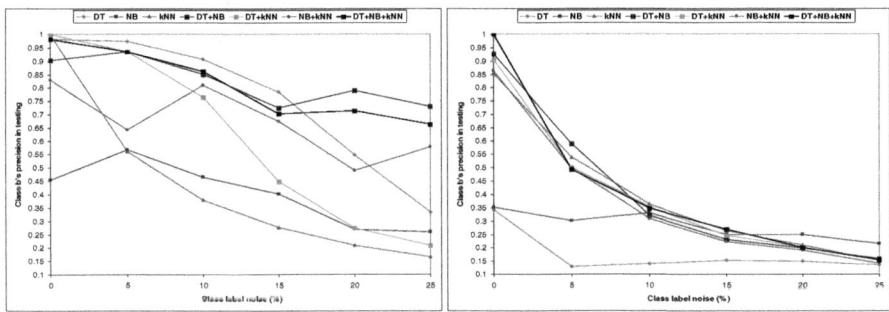

Fig. 14. Precision (minority) vs. noise level for *sick* in bagging (left) and boosting (right).

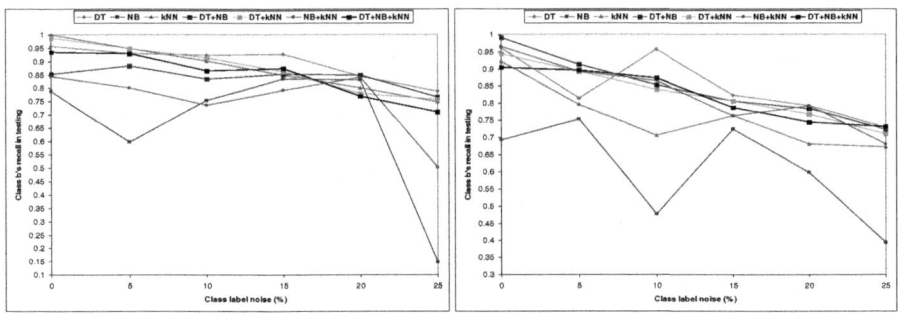

Fig. 15. Recall (minority) vs. noise level for *sick* in bagging (left) and boosting (right)

4) Imbalanced and large data set (*sick*). Results are shown in Fig. 14 and Fig. 15. In terms of precision, (most) curves show similar patterns more in boosting than in bagging; in terms of recall, (most) curves are close to each other more in bagging than in boosting. Moreover, when we mix these three algorithms together, the resulting ensembles (give relatively stable curves, where precision and recall decrease relatively.

5 Conclusions

In the paper, we conduct a rich set of experiments to study the impact of heterogeneity on ensemble, especially focusing on applying bagging and AdaBoost with heterogeneous classifiers to imperfect real-world and synthetic data sets. We report results in the following aspects: 1) diversity values given by homogeneous and heterogeneous classifiers in various measures; 2) classification quality for benchmark data sets that come with mixed types of attributes, missing values, imbalanced class distributions;3) classification quality when data sets are with artificially-added class label noise. Results support that heterogeneity does have impact on the performance of ensemble when the given data sets are not perfect. This paper presents unexpected results and also reveals interesting phenomena that are worth further exploration. Future work include 1) further investigations into the relationship between heterogeneity and classification quality and 2) the study of ensembles composed of heterogeneous classifiers employing techniques that are specifically designed for some types of imperfectness.

References

[1]   Aha, D., Kibler, D.: Instance-based learning algorithms. Mach. Learn. 6(1), 37–66 (1991)
[2]   Aksela, M.: Comparison of Classifier Selection Methods for Improving Committee Performance. MCS, 84–93 (2003)
[3]   Asuncion, A., Newman, D.J.: UCI Machine Learning Repository. School of Information and Computer Science. University of California, Irvine (2007)
[4]   Banfield, R.E., Hall, L.O., Bowyer, K.W., Kegelmeyer, W.P.: Ensemble Diversity Measures and their Application to Thinning. Information Fusion J. 6(1), 49–62 (2005)
[5]   Banfield, R.E., Hall, L.O., Bowyer, K.W., Kegelmeyer, W.P.: A Comparison of De-cision Tree Ensemble Creation Techniques. IEEE Trans. on Pattern Analysis and Machine Intelligence 29(1), 173–180 (2007)
[6]   Banfield, R.E., Bowyer, K.W., Hall, L.O., Kegelmeyer, W.P.: Boosting Lite - Handling Larger data sets and Slower Base Classifiers. MCS (2007)
[7]   Chawla, N., Moore, T., Bowyer, K., Hall, L., Springer, C., Kegelmeyer, P.: Investiga-tion of bagging-like effects and decision trees versus neural nets in protein secondary structure prediction. In: Workshop on Data Mining in Bioinformatics, KDD (2001)
[8]   Chawla, N., Japkowicz, N., Kolcz, A.: Editorial: Special Issue on Learning from Imbalanced data sets. SIGKDD Expl. 6(1), 1–6 (2004)
[9]   Cieslak, D., Chawla, N.: Start Globally, Optimize Locally, Predict Globally: Improving Performance on Unbalanced Data. In: ICDM (2008)
[10] Fan, T.-G., Zhu, Y., Chen, J.-M.: A new measure of classifier diversity in multiple classifier system. ICMLC 1, 18–21 (2008)
[11] Forman, G., Cohen, I.: Learning from Little: Comparison of Classifiers Given Little Training. ECML. HPL-2004-19R1 (2004)
[12] Friedman, J.H.: Multivariate adaptive regression splines. Ann. of Stat. 19(1), 1–67 (1991)
[13] Hettich, S., Bay, S.D.: The UCI KDD Archive. Department of Information and Computer Science. University of California, Irvine (1999)
[14] Hsu, K.-W., Srivastava, J.: Diversity in Combinations of Heterogeneous Classifiers. In: PAKDD (2009)

[15] John, G.H., Langley, P.: Estimating Continuous Distributions in Bayesian Classifiers. In: Conference on Uncertainty in Artificial Intelligence, pp. 338–345 (1995)
[16] Kohonen, T.: Self-Organizing Maps, 3rd edn. Springer Series in Information Sciences, vol. 30 (2001)
[17] Kuncheva, L.I., Whitaker, C.J.: 10 measures of diversity in classifier ensembles: lim-its for two classifiers. In: A DERA/IEE Workshop on Intelligent Sensor Processing, pp. 10/1-10/10 (2001)
[18] Kuncheva, L.I., Whitaker, C.J.: Measures of Diversity in Classifier Ensembles and Their Relationship with the Ensemble Accuracy. Mach. Learn. 51(2), 181–207 (2003)
[19] Kuncheva, L.I.: That elusive diversity in classifier ensembles. In: Perales, F.J., Campilho, A.C., Pérez, N., Sanfeliu, A. (eds.) IbPRIA 2003. LNCS, vol. 2652, pp. 1126–1138. Springer, Heidelberg (2003)
[20] Liu, X.-Y., Wu, J., Zhou, Z.-H.: Exploratory under-sampling for class-imbalance learning. In: ICDM, pp. 965–969 (2006)
[21] Partridge, D., Krzanowski, W.J.: Software diversity: practical statistics for its measurement and exploitation. Information and Software Technology 39, 707–717 (1997)
[22] Quinlan, R.: C4.5: Programs for Machine Learning. Morgan Kaufmann Publishers, San Francisco (1993)
[23] Weiss, G.M.: Mining with Rarity: A Unifying Framework. SIGKDD Expl. 6(1), 7–19 (2004)
[24] Weiss, G.M.: Mining Rare Cases. In: O. Data Mining and Knowledge Discovery Handbook: A Complete Guide for Practitioners and Researchers, pp. 765–776 (2005)
[25] Witten, I.H., Frank, E.: Data Mining: Practical machine learning tools and techniques, 2nd edn. Morgan Kaufmann, San Francisco (2005)
[26] Whitaker, C.J., Kuncheva, L.I.: Examining the relationship between majority vote accuracy and diversity in bagging and boosting, Technical Report, School of Informatics, University of Wales, Bangor (2003)

Undersampling Approach for Imbalanced Training Sets and Induction from Multi-label Text-Categorization Domains

Sareewan Dendamrongvit and Miroslav Kubat

Department of Electrical & Computer Engineering
University of Miami, Coral Gables, FL 33146, USA
s.dendamrongvit@umiami.edu, mkubat@miami.edu

Abstract. Text categorization is an important application domain of multi-label classification where each document can simultaneously belong to more than one class. The most common approach is to address the problem of multi-label examples by inducing a separate binary classifier for each class, and then use these classifiers in parallel. What the information-retrieval community has all but ignored, however, is that such classifiers are almost always induced from highly imbalanced training sets. The study reported in this paper shows how taking this aspect into consideration with a majority-class undersampling we used here can indeed improve classification performance as measured by criteria common in text categorization: macro/micro precision, recall, and F_1. We also show how a slight modification of an older undersampling technique helps further improve the results.

1 Introduction

The classification of text documents under multi-label setting is a necessity in information retrieval. Text categorization organizes text data with the goal of assigning one or more classes from a pre-defined set to a document—for instance, a scientific paper may be labeled as medical studies, clinical research, data analysis and drug tests. We encountered this problem in EUROVOC, a multilingual thesaurus that contains tens of thousands of documents from many different fields. To assist the user's search for a relevant document, the thesaurus needs a powerful indexing scheme, but such scheme is not easy to create. Manual labeling of each single document being impractical due to the enormous size of the collection, the next best solution is to induce some characterization for each class from a training subset of preclassified documents, and then to employ these characterizations to classify the rest of the thesaurus automatically, by a computer program.

Over the past decade, studies of induction from multi-label examples have pursued two fundamental strategies: induction of sets of binary classifiers, and induction of one large multi-label classifier. As for the former, mechanisms based on Bayesian decision theory were studied by [1], [2], and [3], instance-based classifiers were investigated by [4], and the currently popular support vector machines were employed by [5] and [6]. Unfortunately, binary classifiers ignore inter-class relations, which sometimes leads to

T. Theeramunkong et al. (Eds.): PAKDD Workshops 2009, LNAI 5669, pp. 40–52, 2010.

performance degradation. This is why some other researchers preferred to work with multi-label classifiers. Thus [7] modified, accordingly, the methodology of decision trees, while [8] and [9] developed algorithms that handle multi-label domains in the framework of the "boosting" technique invented by [10].

In the study reported here, we concentrated on the first approach: induction of a separate binary classifier for each class. We noticed that previous work had not taken into consideration the fact that some of the binary classifiers have to be induced from imbalanced training sets because negative examples tend to outnumber, significantly, the positive ones. The lack of balance is here almost inevitable. While the total number of classes (in our version of the Eurovoc data) is about thirty, most examples are labeled with no more than four or five classes. We hypothesized that the performance of the induced classifiers would improve if appropriate measures—such as majority-class undersampling—are taken. Moreover, each of the binary classifiers should probably have to rely on a different feature subset.

We focus here on the k-nearest-neighbor classifier (k-NN), selecting relevant features by the use of decision trees. Our idea was that a potential success would inspire other researchers for a closer investigation of other paradigms. We have indeed observed that taking appropriate measures for dealing with imbalanced classes improves performance as measured along the usual text-categorization criteria. Section 2 formally specifies the problem and defines the requisite performance criteria; Section 3 provides details of our approach; and Section 4 details the experiments on two data sets, EUROVOC thesaurus and Reuters corpus. As we were interested in how well our technique works with other paradigms, further experiments with another classifier, Support Vector Machines (SVM), were also carried out.

2   Problem and Performance Criteria

Let $\mathcal{R}^p$ be an instance space, let $\mathcal{X} \subset \mathcal{R}^p$ be a finite set of documents, and let $\mathcal{Y}$ be a finite set of classes such that each $x_i \in \mathcal{X}$ belongs to its subset, $Y_i \subseteq \mathcal{Y}$. The features describing the documents have been obtained from the relative frequencies of words or terms. Given a set of training examples, $S = \{(x_1, Y_1), \ldots, (x_n, Y_n)\}$, the goal is to find a classifier to carry out the mapping $g : \mathcal{X} \rightarrow 2^{\mathcal{Y}}$ in a way that optimizes classification performance. Moreover, the induction of the classifier has to be accomplished in a realistic time.

To obtain reasonable criteria to measure classification performance, let us start with those employed by the field of *information retrieval* for domains where only two class labels are permitted: positive examples and negative examples. Let us denote by TP (true positives) the number of correctly classified positive examples; by FN (false negatives), the number of positive examples misclassified as negative; by FP (false positives), the number of negative examples misclassified as positive ones; and by TN (true negatives), the number of correctly classified negative examples. Let us now use these four quantities to define *precision, Pr,* and *recall, Re,* by the following simple formulas:

$$Pr = \frac{TP}{TP + FP} \qquad\qquad Re = \frac{TP}{TP + FN} \qquad (1)$$

Observing that the user often wants to maximize both criteria, while balancing their values, [11] proposed to combine *precision* and *recall* in a single formula, F_β, parameterized by the user-specified $\beta \in [0, \infty)$ that quantifies the relative importance ascribed to either criterion:

$$F_\beta = \frac{(\beta^2 + 1) \times Pr \times Re}{\beta^2 \times Pr + Re} \qquad (2)$$

It is easy to see that $\beta > 1$ gives more weight to *recall* and $\beta < 1$ gives more weight to *precision*; that F_β converges to *recall* if $\beta \to \infty$, and to *precision* if $\beta = 0$. The situation where *precision* and *recall* are deemed equally relevant is reflected by the value $\beta = 1$, in which case F_1 degenerates to the following formula:

$$F_1 = \frac{2 \times Pr \times Re}{Pr + Re} \qquad (3)$$

Based on these preliminaries, [12] proposed two alternative ways how to generalize these criteria for domains with multi-label examples: (1) *macro-averaging*, where *precision* and *recall* are first computed for each category and then averaged; and (2) *micro-averaging*, where *precision* and *recall* are obtained by summing over all individual decisions. The formulas are summarized in Table 1 where $Pr_i, Re_i, TP_i, FN_i, FP_i$, and TN_i stand for the *precision, recall*, and the four above-mentioned variables for the i-th class.

Table 1. The macro-averaging and micro-averaging versions of the *precision* and *recall* performance criteria for domains with multi-label examples

	Precision	Recall	F_1
Macro	$Pr^M = \frac{\sum_i^k Pr_i}{k}$	$Re^M = \frac{\sum_i^k Re_i}{k}$	$F_1^M = \frac{\sum_i^k F_{1,i}}{k}$
Micro	$Pr^\mu = \frac{\sum_{i=1}^k TP_i}{\sum_{i=1}^k (TP_i + FP_i)}$	$Re^\mu = \frac{\sum_{i=1}^k TP_i}{\sum_{i=1}^k (TP_i + FN_i)}$	$F_1^\mu = \frac{2 \times Pr^\mu \times Re^\mu}{Pr^\mu + Re^\mu}$

For the sake of completeness, we would like to mention that other performance metrics have been recommended (and used) for performance evaluation in multi-label domains, especially in the case of classifiers that make it possible to rank the documents according the the likelihood that they represent the given class. From these, we would like to mention *One_error, Coverage, Average Precision, Hamming loss*, and *Ranking loss* [13,14]. To keep things simple, though, we will not employ them here, although we did recommend their use in our earlier work [9].

3   Proposed Solution

Recall that we want to induce a binary classifier for each class. In the framework of instance-based classifiers (k-NN classifiers), the simplest solution is to store all training examples, described by all available features. This, however, would miss two important points: first, in text categorization, each class is often characterized by different features; second, the training sets for each class are likely to be imbalanced.

As for the first point, we decided to select the features for each class by a decision tree—as advocated, among others, by [15]. For the i-th class, we take the original training set and re-label all examples so that those that have, in their lists of classes, the i-th class are seen as positive, and all others are deemed negative. From this modified training set, we induce (by Quinlan's C4.5 [16] with default parameter settings) a decision tree; the features that appear in the tree are then retained and all others ignored. The result is a modified training set where examples are described by a relevant subset of features, and labeled with "+" or "-" depending on whether or not they belong to the i-th class.

More important for this paper is the second point: in the training set used to induce the i-th classifier, the number of positive examples will often be much smaller than the number of negative examples. We addressed this issue by a minor modification of the technique recommended by [17] who used undersampling of the majority class (the negative examples, in this case). To be more specific, they exploited the idea of Tomek links, informally defined as follows. Take two examples, $\mathbf{x}$ and $\mathbf{y}$, so that each has a different class label. Denote by $\delta(\mathbf{x}, \mathbf{y})$ the distance between $\mathbf{x}$ and $\mathbf{y}$. The pair $(\mathbf{x}, \mathbf{y})$ is a Tomek link if no example $\mathbf{z}$ exists such that $\delta(\mathbf{x}, \mathbf{z}) < \delta(\mathbf{x}, \mathbf{y})$ or $\delta(\mathbf{y}, \mathbf{z}) < \delta(\mathbf{y}, \mathbf{x})$. Examples participating in Tomek links are either borderline or noisy, and as such unreliable. For this reason, the majority-class participant in the Tomek link is removed from the training set.

In our text-categorization domain, we were concerned that this mechanism treats the minority examples as if they were noise-free. However, assuming that some positive examples are incorrectly labeled, it is perhaps inappropriate to remove the (possibly correct) negative examples surrounding them, while retaining the false positive. This conjecture motivated a slight modification of the original idea: if the positive example participating in a Tomek link is "very distant" from any other positive example in the training set (by its Euclidean distance), we "suspect" that it is noisy, and will *not* remove the negative-example part of the Tomek link.

To be more specific, let us denote by $\mathbf{x}$ a positive example participating in a Tomek link, let us denote by $\mathbf{y}$ the the positive example nearest to $\mathbf{x}$, and let $d_{\mathbf{x},\mathbf{y}}$ be the distance between $\mathbf{x}$ and $\mathbf{y}$. Let M be the number of negative examples whose distance from $\mathbf{x}$ is smaller than $d_{\mathbf{x},\mathbf{y}}$. Then, we remove the negative participant of the Tomek link if $M < T$ (where T is a user-specified threshold); otherwise, we do not remove the negative participant.

The pseudocode of the technique thus described is summarized in Table 2. The procedure results in N training sets, each used by a binary classifier representing one class. This means that each binary classifier not only relies on a different set of examples, but these examples are also described by a different feature subset. In the case of k-NN algorithm, when the system is presented with a document, $\mathbf{x}$, to be classified, it will submit the feature vector describing $\mathbf{x}$ to each of the k-NN classifiers in parallel. The i-th k-NN classifier then identifies (among the documents stored in its training set) the k documents that have the smallest Euclidean distance from $\mathbf{x}$. If most of these selected examples are positive, then $\mathbf{x}$ is labeled with the i-class.

Table 2. The pseudocode summarizing our algorithm that creates a different "training set" for each class (to be used by a binary classifier). N denotes the number of class labels.

For $i = 1$ to N:

    1. In the i-th training set, let each example be labeled as positive if the list of its labels contains the i-th class and label it as negative otherwise.

    2. For the i-th training set, induce a decision tree. Only features tested in this tree are deemed relevant. Remove from the i-th training set all irrelevant features.

    3. In the i-th training set, identify all Tomek links. Remove all negative examples participating in the Tomek links, provided that the positive example is not too isolated from the other positive examples.

---

4   Experimental Evaluation

4.1   Data, Methodology, and Parameter Setting

The performance of our technique is demonstrated through experiments on two data sets, a simplified EUROVOC database and a publicly available data set in Reuters test collection, Reuters Corpus Volume 1 version 2 (RCV1-v2) [18].

For *Eurovoc* data, the size of the original database is so huge as to render systematic experimentation impractical: given that each single induction run on the complete data from EUROVOC takes many days, it is impossible to go through the hundreds of experiments needed for statistically justified conclusions. So, as the next-best solution, we decided to work with a simplified database (as we did in our previous paper [9]): 10,000 documents described by 4,000 features, and labeled with only 20 classes. We selected the features by the *Document Frequency* criterion, a method recommended for text categorization by [19]—in principle, we picked randomly 4,000 features from those that appeared in more than 50 documents.

Table 3 summarizes the data, giving for each class the number of examples labeled with it, and then providing the "degree of imbalance" recommended by [20]. Let us denote by n the total number of examples and let us denote by $n_\oplus$ the number of positive examples. The degree of imbalance is then calculated by the following formula:

$$d = |1 - \frac{2n_\oplus}{n}| \tag{4}$$

Note that if 50% examples in a training set belong to one class and 50% to the other, the degree of imbalance is $d = 0$. Conversely, d approaches 1 in the case of highly imbalanced training sets where $n_\oplus \ll n$.

The reader will have noticed that some of the classes are very poorly represented. Intending to use N-fold cross-validation, we decided to ignore those class labels that appear in less than 200 documents because these would be rather underrepresented in the individual "folds." This means that we eliminated the following class labels: 10, 11, 15, 19, 20, 22, 24, 26, 27, and 30.

Table 3. The class-label distribution in Eurovoc data. The degree of imbalance of the training sets used to induce the binary classifiers is calculated by Equation 4.

class	number of representatives	degree of imbalance	class	number of representatives	degree of imbalance
1	1389	0.72	16	1134	0.77
2	2501	0.5	17	2313	0.54
3	4418	0.12	18	2139	0.57
4	1508	0.70	19	5	1
5	1771	0.65	20	2	1
6	3258	0.35	21	1052	0.79
7	1638	0.67	22	1	1
8	2036	0.59	23	423	0.92
9	1393	0.72	24	3	1
10	18	1	25	924	0.82
11	114	0.98	26	0	1
12	1444	0.71	27	1	1
13	1000	0.80	28	4279	0.14
14	935	0.81	29	363	0.93
15	6	1	30	138	0.97

To achieve statistical significance, we used the methodology of 5-fold cross-validation and, whenever appropriate, evaluated the results with t-test. The parameter T (used in the modified Tomek-link technique) was throughout the experiments set to 5% of the size of the training set. For a total of 10,000 examples, and for 5-fold cross-validation, the training set size is 8,000, which means $T = 400$. In auxiliary experiments (not detailed here), we observed that the overall behavior was not very sensitive to the exact setting of T: when we varied T between 2% and 10% of the training set size, the results changed only marginally.

For *RCV1-v2* data, we experimented with five independent training sets and five independent test sets selected from RCV1-v2 collection [18]. Every data set has 3,000 documents described by 4,000 features randomly selected from those that have non-zero values in at least 6 documents. The data are imbalanced with only two classes having more than 500 documents and most classes under 100 documents. The overall degree of imbalance in RCV1-v2 binary sets is higher than in Eurovoc sets shown in Table 3. We used here 53 topic categories that have at least 50 documents, omitting those classes that are extremely imbalanced. For the experiments, the parameter T used in our undersampling technique was set to 0.5% of the size of the training set as it gave the best result when we varied T between 0.1% and 5% of the training set size. The performance measures were evaluated on 5 test data sets.

For both data sets, the values $k \in \{3, 5, 7, 9, 11, 13, 15, 17\}$ for the k-NN classifier were experimented with. A statistical t-test is based on 5% significance level.

---

[1] http://mlkd.csd.auth.gr/multilabel.html#Datasets

4.2   Experiments

We rely on the hypothesis that the imbalanced nature of our training sets impairs the classification performance of the induced binary classifiers; this means that the performance might improve if majority-class undersampling is applied. Our first experiment with Eurovoc data puts this hypothesis to test by comparing the performance obtained in the following scenarios: (1) each k-NN classifier uses all features and all training examples; (2) each k-NN classifier uses all training examples described by features that have appeared in a decision tree induced for this class; (3) each k-NN classifier uses only features that have appeared in a decision tree induced for this class, and only examples that survived the majority-class undersampling mechanism from Section 3.

Table 4. The performance of (1) k-NN, (2) k-NN after feature selection, and (3) k-NN after feature selection and majority-class undersampling on Eurovoc data (we used $k = 9$).

	Micro-averaging			Macro-averaging		
	Precision	Recall	F_1	Precision	Recall	F_1
k-NN	0.49 ± 0.06	0.05 ± 0.01	0.08 ± 0.01	0.67 ± 0.12	0.02 ± 0.00	0.11 ± 0.02
feature+k-NN	0.39 ± 0.02	0.22 ± 0.05	0.28 ± 0.05	0.48 ± 0.04	0.12 ± 0.02	0.19 ± 0.02
feature+sampling+k-NN	0.39 ± 0.01	0.73 ± 0.02	0.51 ± 0.01	0.38 ± 0.02	0.64 ± 0.04	0.46 ± 0.01

Table 4 shows the classification performance (averaged over 5-fold cross-validation) in terms of micro/macro precision, recall, and F_1. In all cases, $k = 9$ was used. The table shows the gradual progression of the performance improvements: first, after the feature selection; then, after subjecting each training set to majority-class undersampling. Most important is of course the increased value of F_1 (which combines precision and recall). A more detailed look reveals that slightly worsened precision is compensated by massive improvement of recall (which was dismal in the case of k-NN). The reader will also have noticed the relatively low standard deviations.

The graphs in Figure 1 show how the performance depends on the concrete value of k (the number of nearest neighbors)—our technique is more robust to variations in k than the other methods. Also, whereas micro-F_1 apparently benefits from the growing value of k, macro-F_1 seems to lose. Which of the two averaging methods really matters depends on the concrete requirements of the given application: whereas micro-F_1 gives equal weight to each class, macro-F_1 weighs the classes according to their relative frequency.

Finally, the graphs in Figure 2 show why we decided to replace the original under-sampling technique from [17] with the improved version from Section 3. While the recall of the original technique is almost 100%, its precision is way too low. What helps is our technique's more careful consideration whether to remove those majority-class examples that seem to surround a noisy positive example. The result is a somewhat lower recall in exchange for better precision and F_1.

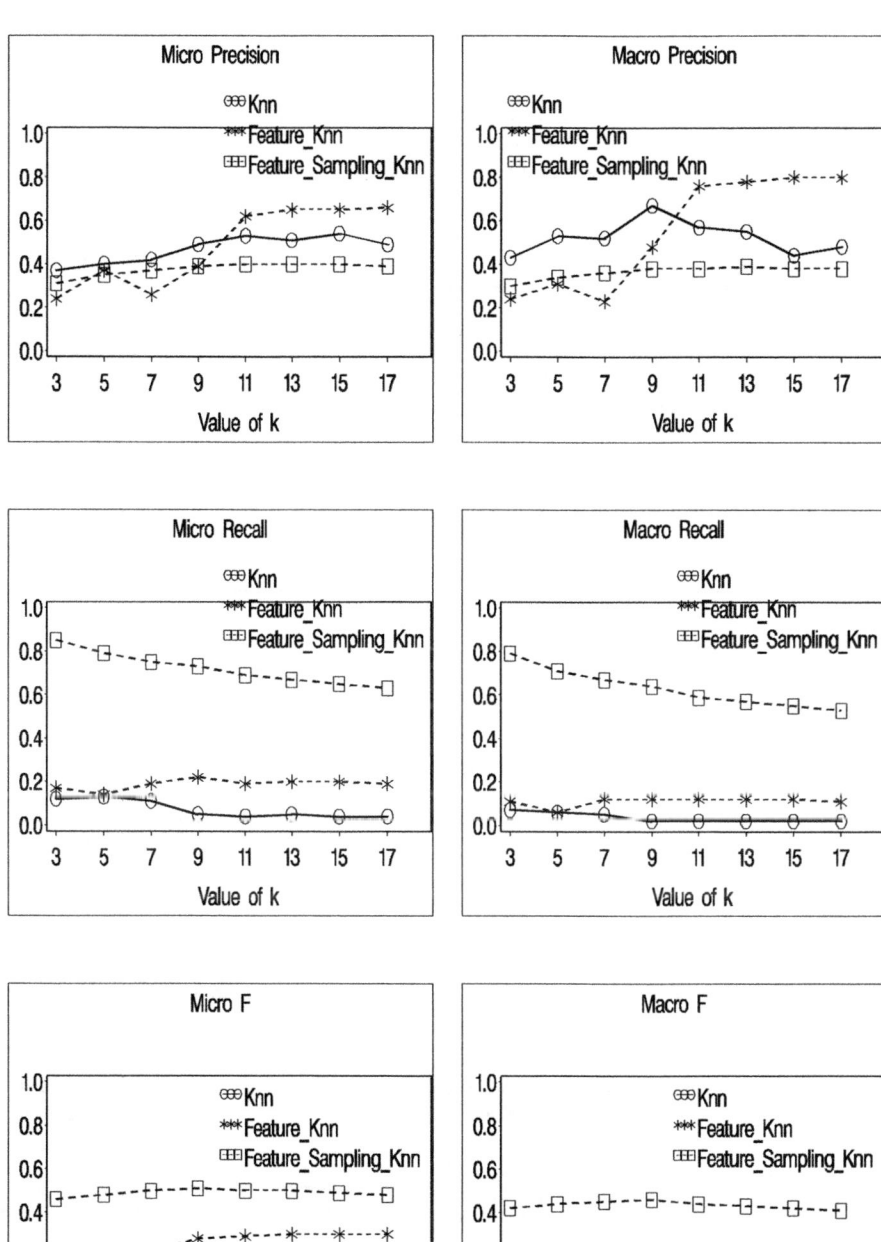

Fig. 1. Sensitivity of our technique to varied number of nearest neighbors, k, on Eurovoc data

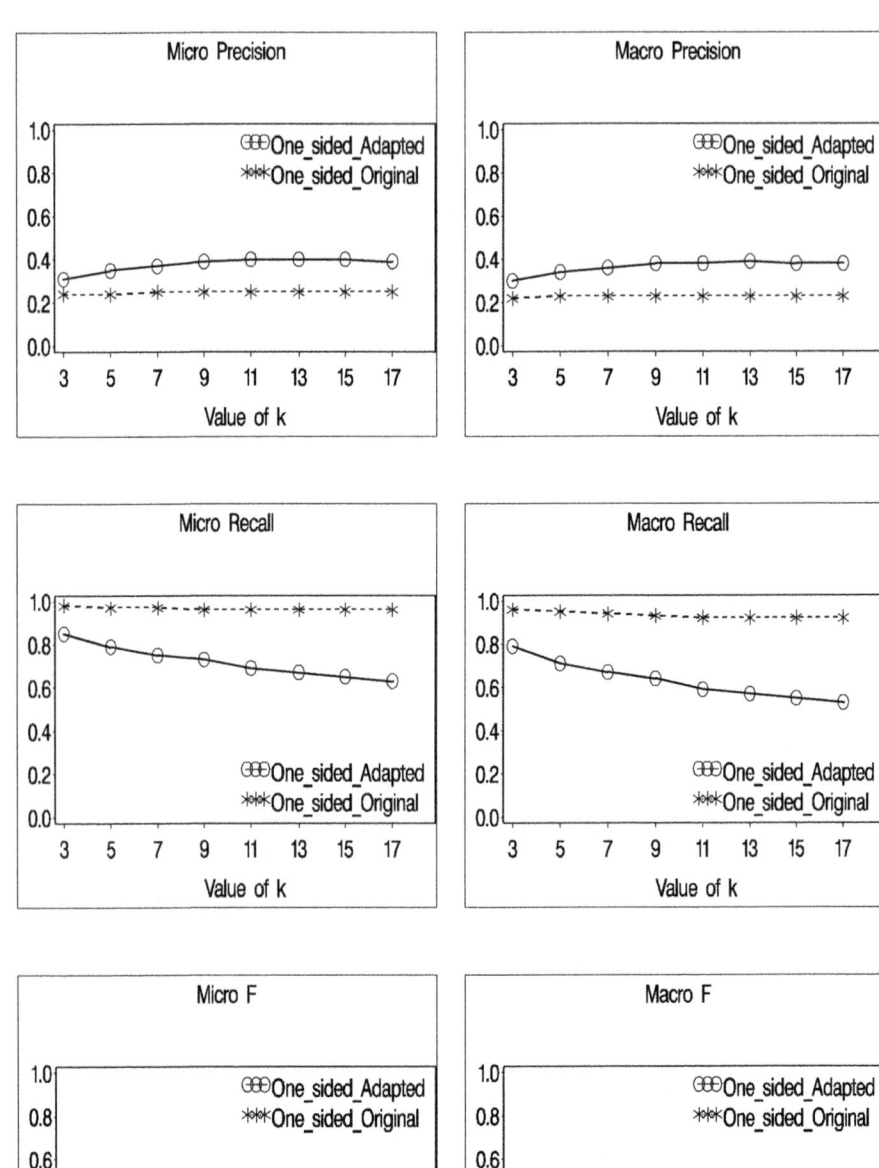

Fig. 2. Comparing the performance of the original algorithm from [17] with the improved version we used here

Table 5. The performance of (1) k-NN, (2) k-NN after feature selection, and (3) k-NN after feature selection and majority-class undersampling on RCV1-v2 data (we used $k = 5$).

	Micro-averaging			Macro-averaging		
	Precision	Recall	F_1	Precision	Recall	F_1
k-NN	0.73 ± 0.14	0.26 ± 0.04	0.38 ± 0.05	0.61 ± 0.05	0.14 ± 0.01	0.18 ± 0.02
feature+k-NN	0.88 ± 0.01	0.44 ± 0.01	0.59 ± 0.01	0.76 ± 0.03	0.33 ± 0.00	0.43 ± 0.01
feature+sampling+k-NN	0.67 ± 0.01	0.65 ± 0.00	0.66 ± 0.01	0.62 ± 0.01	0.49 ± 0.01	0.53 ± 0.01

Having observed improvement in the performance of k-NN classifiers, we are interested to see how our technique performs on other data sets. Here we select a data set from Reuters collection, RCV1-v2, to experiment with. All results from experiments with RCV1-v2 data were obtained as averages from 5 independent test data sets.

The empirical results summarized in Table 5 for RCV1-v2 clearly support the observation drawn from Eurovoc data. Similarly, our technique performs better than other methods in terms of the value of F_1. The improvement in the classification performance after each scenario is similar to what we have observed in Eurovoc data. Note that in this table, $k = 5$ was used in every case.

Figure 3 shows the micro-average and macro-average of three evaluation measures, precision, recall, and F_1, of k-NN algorithm for different k values. The graphs exhibit a similar pattern as in Eurovoc data. Again, our technique outperforms the other methods as it has systematically improved the classification performance along the micro- and macro-averaging versions of F_1.

To see whether the technique helps in other paradigms as well, we experimented also with the Support Vector Machines (SVM). Similarly as before, we denote here (1) svm (2) feature+svm and (3) feature+sampling+svm. Our implementation is based on the LIBSVM [21] with radial basis function. The results in Table 6 and Table 7 reinforce the conclusion from previous experiment: our majority-class undersampling method improves the recall value with acceptable loss in precision. Results from experiments on two data sets agree.

Table 6. The performance of (1) svm, (2) svm after feature selection, and (3) svm after feature selection and majority-class undersampling on Eurovoc data

	Micro-averaging			Macro-averaging		
	Precision	Recall	F_1	Precision	Recall	F_1
svm	0.93 ± 0.02	0.58 ± 0.02	0.72 ± 0.01	0.63 ± 0.01	0.37 ± 0.01	0.46 ± 0.01
feature+svm	0.92 ± 0.01	0.67 ± 0.00	0.78 ± 0.00	0.61 ± 0.01	0.43 ± 0.00	0.50 ± 0.00
feature+sampling+svm	0.87 ± 0.01	0.74 ± 0.02	0.80 ± 0.02	0.57 ± 0.01	0.46 ± 0.01	0.51 ± 0.01

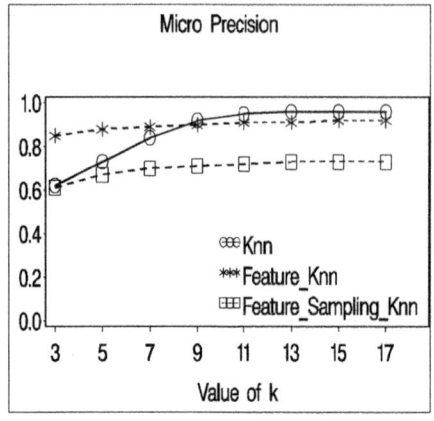
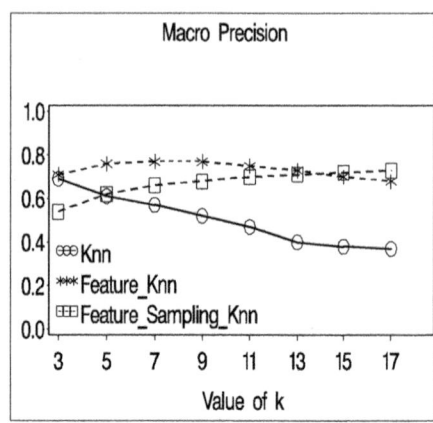
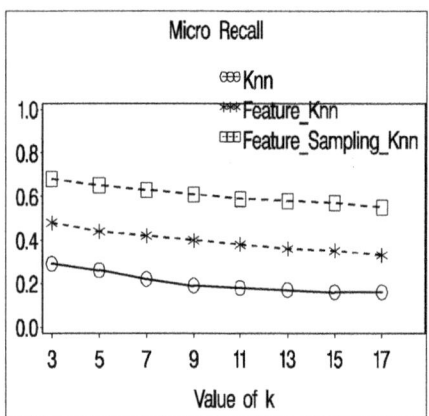
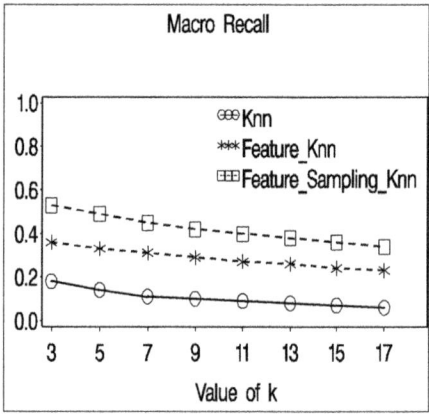
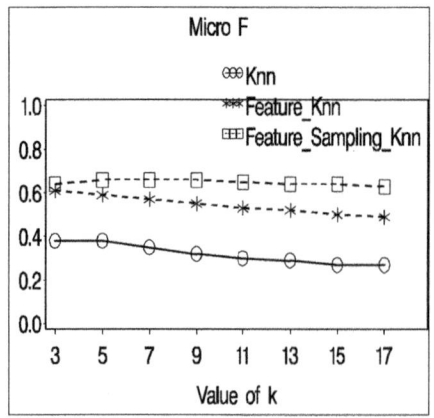
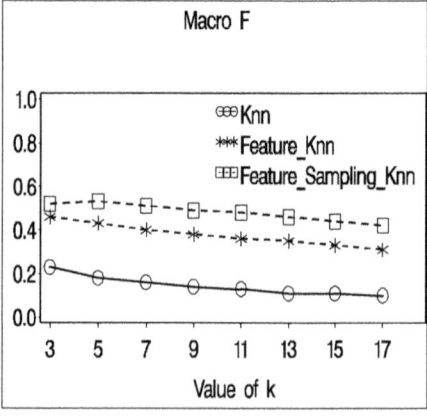

Fig. 3. Sensitivity of our technique to varied number of nearest neighbors, k, on RCV1-v2 data

Table 7. The performance of (1) svm, (2) svm after feature selection, and (3) svm after feature selection and majority-class undersampling on RCV1-v2 data

	Micro-averaging			Macro-averaging		
	Precision	Recall	F_1	Precision	Recall	F_1
svm	0.91 ± 0.02	0.07 ± 0.00	0.12 ± 0.01	0.36 ± 0.04	0.03 ± 0.00	0.05 ± 0.00
feature+svm	0.60 ± 0.01	0.43 ± 0.01	0.50 ± 0.01	0.43 ± 0.01	0.29 ± 0.01	0.33 ± 0.00
feature+sampling+svm	0.57 ± 0.01	0.47 ± 0.01	0.52 ± 0.01	0.41 ± 0.01	0.32 ± 0.00	0.35 ± 0.01

5   Conclusion

The paper reports our experience with the imbalanced-classes treatment in the field of text categorization that is characterized by multi-label examples—each document can potentially have more than one class label at the same time. Classification performance in domains of this kind is typically evaluated along somewhat different criteria—precision, recall, and F_1—than those typical of other machine-learning application.

Having started with the majority-class-undersampling technique from [17], we observed that it indeed improved classification performance; however, we also noticed that the excellent recall had come at the cost of low precision. This observation motivated a modification of the technique, obtaining better performance of the induced binary classifiers. We made the same observation when working with instance-based classifier and when working with Support Vector Machines. Experiments with EUROVOC data showed that our technique improved classification performance for text categorization task. Likewise, further experiments with another data set, documents from the Reuters test collection, confirmed the same outcome.

In summary, the paper has demonstrated that the field of multi-label text-categorization can benefit from explicitly considering the imbalanced-classes problem: while it is common to induce in these domains a separate binary classifier for each class (with the idea of using them all in parallel when classifying future documents), previous studies neglected the fact that each of these binary classifies almost inevitably has to be induced from (sometimes heavily) imbalanced training sets.

Acknowledgment. The research was partly supported by the NSF grant IIS-0513702.

References

1. Langley, P., Iba, W., Thompson, K.: An analysis of Bayesian classifiers. In: Natl. Conf. on Artificial Intelligence, pp. 223–228 (1992)
2. Friedman, N., Geiger, D., Goldszmidt, M.: Bayesian network classifiers. Machine Learning 29(2-3), 131–163 (1997)
3. McCallum, A., Nigam, K.: A comparison of event models for naive Bayes text classification. In: Proc. Workshop on Learning for Text Categorization (AAAI 1998) (1998)

4. Li, B., Lu, Q., Yu, S.: An adaptive k-nearest neighbor text categorization strategy. ACM Trans. on Asian Language Information Processing (TALIP) 3, 215–226 (2004)
5. Joachims, T.: Text categorization with support vector machines: learning with many relevant features. In: Nédellec, C., Rouveirol, C. (eds.) ECML 1998. LNCS, vol. 1398, pp. 137–142. Springer, Heidelberg (1998)
6. Kwok, J.T.: Automated text categorization using support vector machine. In: Proc. Int'l. Conf. on Neural Information Processing (ICONIP 1998), Kitakyushu, JP, pp. 347–351 (1998)
7. Clare, A., King, R.D.: Knowledge discovery in multi-label phenotype data. In: Siebes, A., De Raedt, L. (eds.) PKDD 2001. LNCS (LNAI), vol. 2168, p. 42. Springer, Heidelberg (2001)
8. Schapire, R.E., Singer, Y.: Improved boosting using confidence-rated predictions. Machine Learning 37(3), 297–336 (1999)
9. Sarinnapakorn, K., Kubat, M.: Combining subclassifiers in text categorization: A dst-based solution and a case study. IEEE Transactions on Knowledge and Data Engineering 19(12), 1638–1651 (2007)
10. Schapire, R.E.: The strength of weak learnability. Machine Learning 5(2), 197–227 (1990)
11. van Rijsbergen, C.J.: Information Retrieval, 2nd edn. Butterworths, London (1979)
12. Yang, Y.: An evaluation of statistical approaches to text categorization. Information Retrieval 1(1/2), 69–90 (1999)
13. Schapire, R.E., Singer, Y.: BoosTexter: A boosting-based system for text categorization. Machine Learning 39(2/3), 135–168 (2000)
14. Zhang, M.L., Zhou, Z.H.: A k-nearest neighbor based algorithm for multi-label classification. In: The 1st IEEE Int'l. Conf. on Granular Computing (GrC 2005), Beijing, China, July 2005, vol. 2, pp. 718–721 (2005)
15. Kubat, M., Pfurtscheller, G., Flotzinger, D.: Ai-based approach to automatic sleep classification. Biological Cybernetics 79, 443–448 (1994)
16. Quinlan, J.R.: C4.5: Programs for Machine Learning. Morgan Kaufmann Publishers, San Mateo (1993)
17. Kubat, M., Matwin, S.: Addressing the curse of imbalanced training sets: One-sided selection. In: Proceedings of the 14th International Conference on Machine Learning, ICML 1997, Nashville, TN, pp. 179–186 (1997)
18. Lewis, D.D., Yang, Y., Rose, T.G., Li, F.: Rcv1: A new benchmark collection for text categorization research. Journal of Machine Learning Research 5, 361–397 (2004)
19. Yang, Y., Pedersen, J.O.: A comparative study on feature selection in text categorization. In: Fisher, D.H. (ed.) Proceedings of ICML 1997, 14th International Conference on Machine Learning, Nashville, US, pp. 412–420. Morgan Kaufmann Publishers, San Francisco (1997)
20. Ráez, A.M., López, L.A.U., Steinberger, R.: Adaptive selection of base classifiers in one-against-all learning for large multi-labeled collections. In: Vicedo, J.L., Martínez-Barco, P., Muñoz, R., Saiz Noeda, M. (eds.) EsTAL 2004. LNCS (LNAI), vol. 3230, pp. 1–12. Springer, Heidelberg (2004)
21. Chang, C.C., Lin, C.J.: LIBSVM: a library for support vector machines (2001), Software available at http://www.csie.ntu.edu.tw/~cjlin/libsvm

Adaptive Methods for Classification in Arbitrarily Imbalanced and Drifting Data Streams

Ryan N. Lichtenwalter and Nitesh V. Chawla

The University of Notre Dame, Notre Dame, IN 46556, USA

Abstract. Streaming data is pervasive in a multitude of data mining applications. One fundamental problem in the task of mining streaming data is distributional drift over time. Streams may also exhibit high and varying degrees of class imbalance, which can further complicate the task. In scenarios like these, class imbalance is particularly difficult to overcome and has not been as thoroughly studied. In this paper, we comprehensively consider the issues of changing distributions in conjunction with high degrees of class imbalance in streaming data. We propose new approaches based on distributional divergence and meta-classification that improve several performance metrics often applied in the study of imbalanced classification. We also propose a new distance measure for detecting distributional drift and examine its utility in weighting ensemble base classifiers. We employ a sequential validation framework, which we believe is the most meaningful option in the context of streaming imbalanced data.

Keywords: Sequential learning, stream mining, imbalanced data, skew, concept drift, Hellinger distance.

1 Introduction

Data stream mining is a prolific area of research. In recent years, many papers have been published either directly related to stream mining or addressing challenges in a particular stream mining application. Concept drift is a problem inherent in stream mining that continues to limit performance. Classifier ensembles have been applied in several incarnations to reduce variance, but they cannot combat bias. More recently the problem of class imbalance has been considered in the context of existing stream mining research. The intersection of these difficulties makes for a uniquely interesting and demanding research topic that can contribute to practically every data-driven or data-related field. As data streams become increasingly ubiquitous and prolific, the importance of solving their unique and interrelated challenges grows.

Research in stream mining primarily falls into one of two families: incremental processing [1,2,3,4] or batch processing [5,6]. Figure 1 illustrates. In the former approach, each processing step handles one instance. Labels for an instance are

T. Theeramunkong et al. (Eds.): PAKDD Workshops 2009, LNAI 5669, pp. 53–75, 2010.

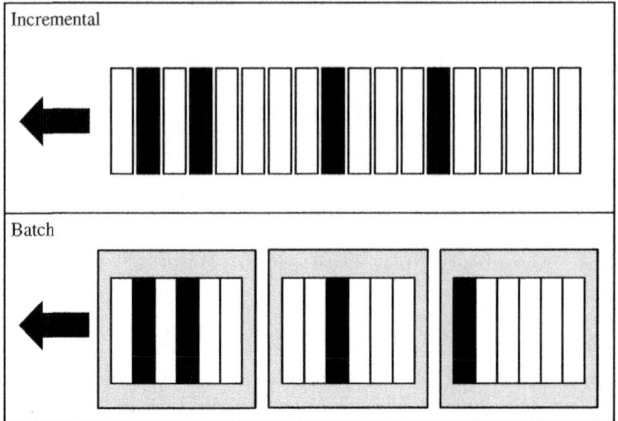

Fig. 1. An illustration of the difference in the way incremental and batch processing methods expect and handle incoming streams of data. With the arrival of each subsequent unit, class labels become available for the previous unit.

assumed to be available immediately after classifying the instance. Instance processing methods have an inherent advantage in terms of their adaptiveness due to this assumption. At time t, instance i_t is received and classified with some model trained on instances i_1 through i_{t-1}. At time $t+1$, a label arrives for instance i_t and unlabeled instance i_{t+1} becomes available for classification. In the latter approach, each batch is a collection of data instances that arrives during a particular period. At time t, batch b_t is received and classified with some model trained on batches b_1 through b_{t-1}. At time $t+1$, labels arrive for batch b_t and unlabeled instances in batch b_{t+1} become available for classification. While the problem spaces for these two approaches do intersect, often the instance processing assumption is unrealistic. In many scenarios, large sets of new data and class labels arrive with low temporal granularity such as weeks, months, or years. The principles presented in this paper operate in the batch processing space.

2   Challenges in Data Streams

Stream mining presents inherent difficulties not present in static distributions. Most significantly, since data streams are generated over time by an underlying hidden function, changes in that function can cause various forms of data drift. Secondarily, the rare class problem becomes more difficult because positive class events, which appear infrequently in a static distribution, may be available in even smaller quantities per unit time in data streams.

2.1   Distributional and Concept Drift

Since data streams are generated over time by an underlying hidden function, changes in that function can cause changes in data distributions and posterior probabilities, a phenomenon termed *concept drift*. Concept drift occurs when any of the following are present: $\Delta P(\boldsymbol{f})$, a change in the distribution of feature values; $\Delta P(c|\boldsymbol{f})$, a change in the conditional probability of class labels for particular feature values; and both $\Delta P(c|\boldsymbol{f})$ and $\Delta P(\boldsymbol{f})$, when both the feature distribution and the conditional probability of class label appearance change. Examples of these three forms of drift are illustrated in Figure 2.

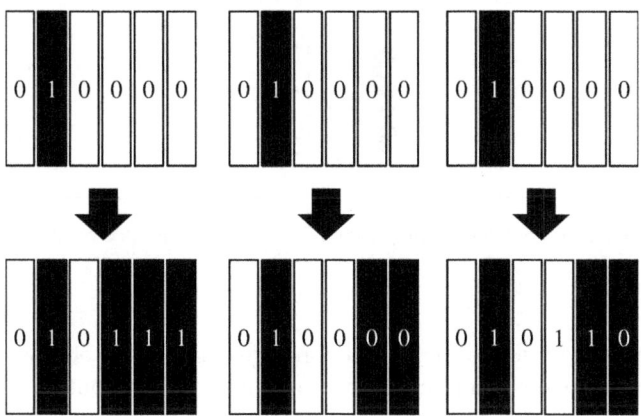

Fig. 2. An illustration of the three different forms of concept drift. The class value is denoted by colors and the feature value is denoted by numbers.

- $\Delta P(\boldsymbol{f})$ (left):
  - $P(0) = \frac{5}{6} \rightarrow P(0) = \frac{1}{3}$
  - $P(1) = \frac{1}{6} \rightarrow P(1) = \frac{2}{3}$
- $\Delta P(c|\boldsymbol{f})$ (middle):
  - $P(white|0) = 1 \rightarrow P(white|0) = \frac{3}{5}$
  - $P(black|0) = 0 \rightarrow P(black|0) = \frac{2}{5}$
- $\Delta P(c|\boldsymbol{f})$ and $\Delta P(\boldsymbol{f})$ (right):
  - $P(0) = \frac{5}{6} \rightarrow P(0) = \frac{1}{2}$
  - $P(1) = \frac{1}{6} \rightarrow P(1) = \frac{1}{2}$
  - $P(white|0) = 1 \rightarrow P(white|0) = \frac{2}{3}$
  - $P(white|1) = 0 \rightarrow P(white|1) = \frac{1}{3}$
  - $P(black|0) = 0 \rightarrow P(black|0) = \frac{1}{3}$
  - $P(black|1) = 1 \rightarrow P(black|1) = \frac{2}{3}$

Henceforth, we designate changes that occur solely in feature distributions, $\Delta P(\boldsymbol{f})$, with the term *distributional drift*.

2.2 Class Imbalance

Imbalanced data is a persistent problem in general data mining contexts. The problem becomes especially difficult in streams. Consider data with 12,000 instances and a class ratio of 100:1. This leaves 120 positive class examples. This

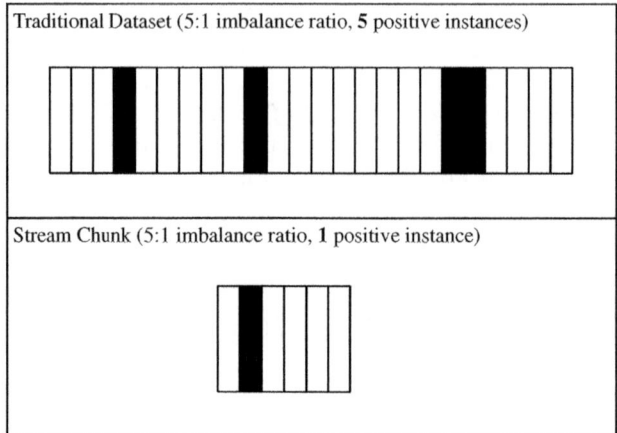

Fig. 3. An illustration of the manifestation of the class imbalance problem in the stream mining domain

may be sufficient to define the class boundaries and leaves plenty of data after application of a good resampling approach. In a streaming context this data may appear over the course of a year. If we assume a uniform temporal distribution of the positive class and want a good model for monthly predictions, we have only 10 positive class examples each month on which to train. This may be insufficient. In reality, some months may exhibit no positive class occurrences at all while others have hundreds. Figure 3 illustrates a simplistic example.

2.3   Problem Intersections

When imbalanced data and concept drift are both present in a data stream, they can cause several confounding effects. First, class frequency inversions for a class c with positive value c_i and negative value c_j can occur such that prior probabilities $P(c_i) << P(c_j)$ may become $P(c_i) >> P(c_j)$ later in the stream. An effective model must be able to adapt to changes of these types. Furthermore, for a given decision threshold θ, posterior probabilities may change for the positive class such that $P(c_i|\boldsymbol{f}) >> \theta$ may become $P(c_i|\boldsymbol{f}) << \theta$. The lack of prevalence of the positive class may result in difficulty assimilating the new concept, thereby rendering models unable to detect the class effectively for an extended period of time.

3   Evaluation Methods

There are two fundamental methods of evaluating classification performance in data streams under the batch paradigm. The first is to evaluate the classifier over a single batch among the many batches of data. The second is to evaluate the classifier over all of the batches in the data stream and either examine them individually or use some notion of average performance. We highlight the core ideas underlying these two fundamental methods.

Evaluating the classifier over a single batch of data has been used in [5], where the last batch of data is used to render performance scores. Such an evaluation framework might somehow seek to report a concept of average classification accuracy by fulfilling some combination of these three goals:

1. Reporting results on a batch of data representative of the stream as a whole.
2. Reporting results on a batch of data for which classifier performance is representative of some notion of typical or average performance.
3. Reporting results on a batch of data sufficiently late in the stream so that the classifier has captured the concept.

The first and second may seem at first to be the same, but in fact they are not. A simple example proves this. Suppose a heavily drifting binary classification problem C with batches b_1 to b_n and class labels $\in \{0, 1\}$. Further suppose a dumb model M that always adapts to drift by classifying all instances of b_i with the predominant label from b_{i-1}. Now suppose that batches b_1 to b_{n-1} alternate such that instances $\in \{b_i | i = 2m, m \in \mathbb{Z}\}$ have label 0, instances $\in \{b_i | i = 2m + 1, m \in \mathbb{Z}\}$ have label 1, and b_n has a uniform distribution of the two classes. When M is evaluated according to the first goal, b_n is used and M is reported to obtain 50 percent classification accuracy. When M is evaluated according to the second goal, any one of b_1 through b_{n-1} is used and M is reported to obtain 0 percent classification accuracy.

The example is contrived, but nonetheless illustrates that the first two goals are not always aligned in terms of the results they output. Unfortunately, the first goal can output virtually meaningless results and the second goal may be difficult to achieve without evaluating model performance on all batches anyhow. The result is that it may be very difficult to decide on which batch one should evaluate and report performance, especially in the context of the third goal.

The third goal itself may or may not be desirable. Since one of the principle objectives in many approaches to data stream mining is to overcome various forms of drift, waiting until the model has achieved a stable state with respect to its classification performance in the stream may inflate the performance metric. Even in data streams with static distributions and concepts, using a single batch such as the last one does not include information about how quickly the target concept was learned. Two models M_1 and M_2 that reach performance level P somewhere in n batches of data certainly are not equally effective if M_1 reaches P after $\frac{n}{2}$ batches and M_2 reaches P after $\frac{n}{4}$ batches. The second model, in most circumstances should be considered clearly superior, but evaluation on the last batch will consider them the same.

All of these problems may be avoided by employing the second fundamental method. Using all batches does not suffer from the problem of finding representative data. Indeed, in data streams where concepts fluctuate through time, the existence of any meaningful sense of representative data is dubious since a single batch fundamentally lacks the ability to represent the temporal dimension. Using all batches also guarantees that not only the peak performance is captured in the evaluation, but also the speed at which that performance is reached. For these reasons, we advocate, and unless otherwise stated, employ the evaluation method that averages classification performance metrics over all batches in the data stream.

4   Boundary Definition

We present two findings about how to resample in stream mining to achieve better performance under heavy class imbalance. The first, in line with other findings such as in [7], is that the performance improvement in propagating rare-class instances arises from defining the class boundary better rather than simply providing more positive class instances on which to train. The second arises from the first. In addition to propagating rare-class instances, we propagate instances in the negative class that the current model misclassifies. These help further define the boundary thus increasing precision while minimally affecting recall.

5   Data Set Distance

It is desirable to adapt to drift in a data stream as quickly as possible. Adjusting ensemble weights based on classification performance metrics is a common method, but this is always one batch behind. In a worst case scenario, the data stream fluctuates such that performance on batch i and performance on batch $i+1$ are inversely correlated. Applying weights based on performance metrics in such a scenario will perform worse than not using any weighting scheme at all. We can do better by examining distributional and information-theoretic properties of the data in the past and comparing the results to the testing batch.

5.1   Hellinger Distance

Hellinger distance has recently been used with excellent effect in detecting classifier performance degradation due to distributional changes [8] and even as a decision tree splitting criterion [9]. For this reason, we employ it to construct a consistent measure of distance between two separate data sets or data batches. After discretizing numeric attributes into some number of equal-width bins, we can define Hellinger distance in two batches X and Y for a given feature f as:

$$HD(X, Y, f) = \sqrt{\sum_{v \in f} \left(\sqrt{\frac{|X_{f=v}|}{|X|}} - \sqrt{\frac{|Y_{f=v}|}{|Y|}} \right)^2} \tag{1}$$

Algorithm 1. Misclassified Propagation

Require: unlabeled batch of instances b
    model m
    real class labels l
    desired percent positive p
    set of positive examples P
    set of misclassified negative examples N
Ensure: resampled batch b
    updated set of positive examples P
    updated set of misclassified negative examples N

1: **for** $instance \in b$ **do**
2:    $label \leftarrow \text{PREDICT}(m, instance)$
3:    **if** $l_{instance} = +$ **then**
4:       $pos \leftarrow pos \cup instance$
5:    **else**
6:       $neg \leftarrow neg \cup instance$
7:       **if** $l_{instance} \neq label$ **then**
8:          $mneg \leftarrow mneg \cup instance$
9:       **end if**
10:   **end if**
11: **end for**
12: **while** $|pos| < p * (|pos| + |neg| + |mneg|)$ **do**
13:   **if** $|neg| > 0$ **then**
14:      REMOVE-RANDOM(neg)
15:   **else**
16:      REMOVE-RANDOM($mneg$)
17:   **end if**
18: **end while**
19: $b \leftarrow pos \cup neg \cup mneg$
20: **return** b, P, N

5.2 Information Gain

To extend Hellinger distance to measure the distance between two data sets, we must first devise some method of accounting for the difference in feature relevance. Simple aggregating functions fail to capture this important information. For two features f_i and f_j in training sets X and Y and testing set Z, it is possible that $HD(X, Z, f_i) + HD(X, Z, f_j) << HD(Y, Z, f_i) + HD(Y, Z, f_j)$ and the model trained on X performs more poorly than the model trained on Y. This can happen when $HD(X, Z, f_i) - HD(Y, Z, f_i) < HD(X, Z, f_j) - HD(Y, Z, f_j)$ and f_i is a more relevant feature. In other words, the two distributions of instances for f_i are more proximate than the two distributions of instances for f_j, but f_j is a weaker feature. If we assume that feature relevance remains roughly stable over time we can use the information gain on labeled data to inform the Hellinger distance. Information gain for a batch X is defined as the decrease in entropy H of a class c conditioned upon a particular feature f.

$$IG(X, f) = H(X_c) - H(X_c|X_f) \tag{2}$$

We observe immediately that the product of information gain and Hellinger distance behaves well for most ordinary situations. Because both Hellinger distance and information gain are well-defined, their product is also well-defined with the range $[0, \sqrt{2}]$ and has a sensible meaning. When information gain is high and Hellinger distance is high, the product is high, corresponding to when a highly discriminating feature has significantly drifted. When information gain is high and Hellinger distance is low, the product is low, corresponding to when a highly discriminating feature has remained stable. When information gain is low and Hellinger distance is high, the product is, unfortunately, low.

In this final case, problems may arise. Consider the most extreme case of a batch in which there are no rare class instances. In these cases, there is no reduction in entropy by conditioning on any feature because the initial class distribution has no entropy. The judged difference between this batch and any other for the feature is 0 no matter how divergent the feature distributions are. In reality, our distance function should fall back on Hellinger distance. This way it would not consider two data batches to have 0 distance just because the training set has no rare-class instances and the otherwise useless "predict 1" model is only applied when Hellinger distance indicates that it should be. We can easily achieve this with a simple smoothing, leading us to the following distance function for a single feature.

$$HDIG(X, Y, f) = HD(X, Y, f) * (1 + IG(X, f)) \tag{3}$$

To obtain a normalized distance between two complete data sets, we can apply an arbitrary aggregation function to these relevance-weighted feature distribution distances. For our experiments, we use summation and arrive with the following distance function for two data sets X and Y, which is well defined in the range $[0, 2\sqrt{2}]$ for all data sets:

$$D(X, Y) = \frac{\sum_{f \in X} HDIG(X, Y, f)}{|f \in X|} \tag{4}$$

We apply algorithm 2 to incoming batches to classify their instances with distributions that were most similar in the past.

A distance function measuring distance between data sets X and Y need not be correlated with the performance on Y of a model trained on X for it to be logical, consistent, and useful. For this particular application, however, such a correlation is indicative of a strong potential to provide good weights. We illustrate in Figure 4 a plot of the correlation between the distance and the F_1-measure that results from weighting the ensemble according to the distance. Pearson correlation coefficient provides a rough single number representing the effectiveness of the distance function. For the illustrated example, the coefficient is 0.435. More important is the performance of the function when distances vary greatly. Ignoring the noise in the lower right quadrant of the plot, we observe substantial performance differences correspond to substantial distance differences.

Algorithm 2. Distance-Weighted Ensembles

Require: batches $b_1 \ldots b_t$
    models $m_1 \ldots m_t$
    unlabeled batch b_{t+1}
Ensure: Probability distributions P for instances $\in b_{t+1}$

1: **for** $i = 1$ to t **do**
2:    $distance_i \leftarrow D(b_i, b_{t+1})$
3:    $weight_i \leftarrow \frac{1}{distance_i}$
4: **end for**
5: NORMALIZE(w)
6: **for** $instance \in b_{t+1}$ **do**
7:    **for** $i = 1$ to t **do**
8:       $P_{instance} \leftarrow P_{instance} + w_i \cdot \text{PREDICT}(m_i, instance)$
9:    **end for**
10: **end for**
11: **return** P

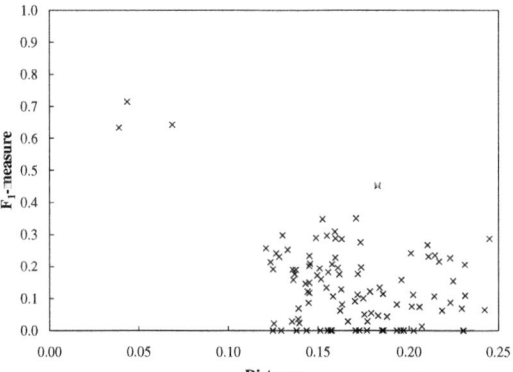

Fig. 4. The correlation between HD-IG distance output and weighted ensemble classification performance. This plot is computed based on the `compustat` data set.

6   Meta-classification

The product of Hellinger distance and information gain ensures that distances appropriately reflect influential features. There are two caveats: (1) it is still possible that the influence of $\Delta P(c|\boldsymbol{f})$ is unrelated to or overpowers $\Delta P(\boldsymbol{f})$ and (2) the distances are constructed respecting the importance of the most powerful features only as measured by the training batch. Feature relevance may change over time, which poses a problem when assigning weights to old batches based on properties such as information gain. Without class labels, it is impossible to determine with complete certainty the relevance of a feature in testing data.

Algorithm 3. Ensembles Weighted Using Meta-Classification

Require: ordered labeled batches $b_1 \ldots b_t$
    models m_1 to m_t
    unlabeled batch b_{t+1}
    regression classifier R
    list of performance features $\boldsymbol{F}$
    objective performance metric f_o
Ensure: probability distributions $\boldsymbol{P}$ for instances $\in b_{t+1}$
    updated regression classifier R

  1: TRAIN(R)
  2: **for** $i \leftarrow 1$ to n **do**
  3:    $w_i \leftarrow$ PREDICT(R, b_i)
  4: **end for**
  5: NORMALIZE($weights$)
  6: **for** $instance \in b_{t+1}$ **do**
  7:   **for** $i = 1$ to t **do**
  8:     $\boldsymbol{P_{instance}} \leftarrow \boldsymbol{P_{instance}} + w_i \cdot$ PREDICT($m_i, instance$)
  9:   **end for**
10:   **for** $feature \in b_{t+1}$ **do**
11:     ADD-FEATURE($meta, HD(b_i, b_{t+1}, feature)$)
12:   **end for**
13:   **for** $f \in F$ **do**
14:     ADD-FEATURE($meta,$ EVALUATE-METRIC($P_{instance}, f$))
15:   **end for**
16:   ADD-CLASS($meta,$ EVALUATE-METRIC($P_{instance}, f_o$))
17:   ADD-INSTANCE($R, \boldsymbol{meta}$)
18: **end for**
19: **return** $\boldsymbol{P}, R$

We present another method of incorporating distributional divergence measures into ensemble model weighting that can respond to posterior probability drift. We propose a meta-level weighting classifier with feature vector $HD(X, Y, f) \forall f \in \boldsymbol{f}$, and a sliding window of classifier performance metrics. The Hellinger distance metrics provide information about distributional divergence at the level of individual features while the performance metrics allow for weighting based on strong trends in classification success patterns. The class is the performance metric to optimize. After each base classifier makes its prediction and receives performance results, whether the prediction contributes to the ensemble or not, a new instance is generated in the weighting classifier data set. The weighting classifier must be rebuilt and and must then perform regression analysis on the features. In our experiments, we use simple linear regression. Algorithm 3 delineates the process in its most general form. There are many possible modifications such as using only a subset of best base classifiers and modifying frequency with which information is added to the meta-classifier.

Table 1. The format of the data on which regression is performed to predict performance. The example represents a single instance of the metadata containing information about a model trained on b_i and previously used to classify b_j.

Weighting Classifier Features		Class
Distributional	Performance	Objective Metric
$HD(b_i, b_{t+1}, f) \forall f \in b_i$	$MSE(b_j)$	$AUC\text{-}ROC(b_{t+1})$

We leave these for future consideration. The algorithm does not provide the mundane details of training the base classifier, but instead assumes that this is performed separately. In Table 1 we provide an illustration of the weighting classifier data format with AUC-ROC as the optimization objective:

In the same spirit as Section 5.2, we provide Figure 5 to quantify the effectiveness of the meta-classification scheme. The correlation coefficient is 0.577 for F_1-measure and a similar value for other performance measures. This particular example depicts the ability of the meta-classification scheme to handle drift in posterior probabilities alone.

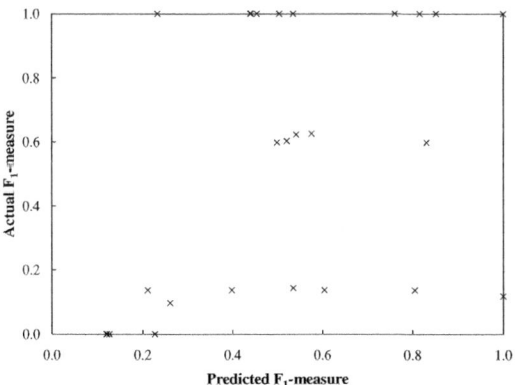

Fig. 5. The correlation between the predicted performance and the actual weighted ensemble classification performance. This plot is computed based on the STAGGER data set.

7   Weighting

Even after rendering a distance performance prediction, it is necessary to apply some function to transform outputs to weights. In the case of a distance d, weight should be inversely proportional to closeness, suggesting a function such as $\frac{1}{d}$. In the case of most performance metrics, for which higher values are better, the most simple function given a prediction value p is to assign p directly as the weight.

There are two problems with using such simple functions generally: (1) differences in the values may be too insignificant to create discriminating weights and (2) many bad models can outweigh few good ones. At the same time, we have to be careful not to lose the effectiveness of the ensemble at reducing variance in cases where differences in the base classifiers are insubstantial. We can attack the first problem by using more complex functions such as exponential or power functions. For a distance d our weight would then be x^{-d} or d^{-x} using some base or exponent x. Given a sufficiently large number of batches, any such function, no matter what disparity it creates in the model weights, will allow possibly much better classifiers to be out-ruled. To combat this, we can incorporate the number of ensemble members b to give us functions such as d^{-b} or d^{-bx}.

While these considerations are important, the require a completely separate study involving many data sets and extensive parameter exploration. Inevitably there will still be some small number of classification problems that defy any single general function. For these reasons, we acknowledge this difficult issue, but sidestep it and apply linear weighting functions to achieve all of our results.

8   Data

We operate on different categories of data sets with drastically different properties. Chief among these in number are UCI data sets [10]. They are not inherently sequential and exhibit no concept drift, but we consider these data sets because of their class imbalance. We render them as data streams by randomizing the order of instances and processing them in batches. Most notably, we use `thyroid`, `covtype`, `optdigits`, and `letter-recognition` for direct comparison with [5].

We also present several generated and real world data sets that vary greatly in their degrees of imbalance and distributional drift. The `can` data set captures properties of a crunching can over time and exhibits imbalance at a ratio of 52:1 and extreme distributional drift in the form of blips of positive class instances. `compustat` includes both class imbalance and concept drift, as the training and testing sets represent several years of changing data. The `football` data set comprises features derived from 2003-2008 college football statistics available on ESPN and the class is whether the home team wins. Features drift over time due to underlying changes, the most significant of which are clock rule changes that affect game time. We use a version of the `kddcup2008` data set, which contains instances of breast cancer and exhibits extreme imbalance. `text` is a text-recognition data set with a large number of features. Finally, we include an ordered variation of the Wharton `wrds` data set.

Characteristics of all of these data sets are available in Table 2. Lack of public availability of source data is a problem with many works focusing on streaming data. It is therefore difficult to repeat or verify experiments. For this reason, all data sets for experiments in this paper are publicly available.

Table 2. Data Set Characteristics

Name	Instances All	Instances Positive	Features Nominal	Features Numeric	Properties Imbalance	Properties Ordered
adult	48,842	11,687	8	6	3.2:1	No
boundary	3,505	123	175	0	27.5:1	No
breast-w	569	212	0	30	1.7:1	No
cam	18,916	942	132	0	19.1:1	No
can	443,872	8,360	0	9	52.1:1	Yes
compustat	13,657	520	0	20	25.3:1	Yes
covtype	38,500	2,747	0	10	13.0:1	No
estate	5,322	636	0	12	7.4:1	No
football	4,288	1,597	2	11	1.7:1	Yes
fourclass	862	307	0	2	1.8:1	No
german	1,000	300	0	24	2.3:1	No
ism	11,180	260	0	6	42.0:1	No
kddcup2008	102,294	623	2	123	163.2:1	No
letter	20,000	789	0	16	24.3:1	No
oil	937	41	0	49	21.9:1	No
ozone-1h	2,536	73	0	72	33.7:1	Yes
ozone-8h	2,534	160	0	72	14.8:1	Yes
page	5,473	560	0	10	8.8:1	No
pendigits	10,992	1,142	0	16	8.6:1	No
phoneme	5,400	1,584	0	5	2.4:1	No
phoss	11,411	613	0	480	17.6:1	No
pima	768	268	0	8	1.9:1	No
satimage	6,430	625	0	36	9.3:1	No
segment	2,310	330	0	19	6.0:1	No
splice	1,000	483	0	60	1.1:1	No
STAGGER	12,000	5,303	3	0	1.3:1	Yes
svmguide1	3,089	1,089	0	4	1.8:1	No
text	11,162	709	0	11,465	14.7:1	Yes
thyroid	7,200	166	15	6	42.4:1	No
wrds	99,200	48,832	2	39	1.0:1	Yes

9 Results

We first examine the degree to which the boundary definition (BD) method can improve classification metrics on static data sets. Because this method seeks only to improve performance in the domain of extreme imbalance, we compare it to the other existing work [5], from which it borrows the idea of propagating minority class instances. In Tables 3(a), 3(b), and 3(c) we directly compare the performance for several data sets reported in [5]. We duplicated their methods (SE) for constructing the data sets and evaluating performance as closely as possible. For letter we created a separate data for each individual letter with that letter as the positive class. For optdigits we followed the same process for each number. We select class 2 as the negative class and class 4 as the positive class in covtype. From thyroid we create two data sets with class 3 as the

Table 3. Direct comparison of BD and SE

(a) Precision and Recall

	Precision				Recall			
	BL	SE-	SE	BD	BL	SE-	SE	BD
covtype	0.980	0.603	0.770	**0.993**	0.958	**0.999**	0.998	**0.999**
letter	**0.836**	0.398	0.519	0.780	0.719	**0.963**	0.955	0.934
optdigits	0.869	0.765	0.834	**0.915**	0.827	**0.954**	0.952	0.952
thyroid1	**0.914**	0.527	0.728	0.897	0.907	**1.000**	**1.000**	**1.000**
thyroid2	0.940	0.617	0.857	**0.957**	0.982	**1.000**	**1.000**	**1.000**

(b) F_1-measure and PRBEP

	F_1-measure				PRBEP			
	BL	SE-	SE	BD	BL	SE-	SE	BD
covtype	0.969	0.748	0.869	**0.996**	0.968	0.880	0.953	**0.997**
letter	0.771	0.558	0.667	**0.847**	0.739	0.775	0.826	**0.891**
optdigits	0.846	0.847	0.887	**0.932**	0.833	0.896	0.915	**0.936**
thyroid1	0.910	0.686	0.836	**0.945**	0.899	0.876	0.912	**0.937**
thyroid2	0.960	0.762	0.922	**0.978**	0.941	0.860	0.922	**0.975**

(c) AUC-ROC and AUC-PR

	AUC-ROC				AUC-PR			
	BL	SE-	SE	BD	BL	SE-	SE	BD
covtype	0.975	0.999	**1.000**	**1.000**	0.944	0.840	0.951	**1.000**
letter	0.891	0.987	0.989	**0.993**	0.675	0.785	0.847	**0.941**
optdigits	0.909	0.990	0.990	**0.993**	0.768	0.934	0.949	**0.972**
thyroid1	0.951	0.998	**0.999**	**0.999**	0.854	0.897	0.913	**0.934**
thyroid2	0.991	0.995	0.998	**0.999**	0.934	0.862	0.920	**0.975**

negative class: one with 1 as the positive class and the other with 2 as the positive class. After constructing the data sets from the UCI source data, we randomized the order of the instances and repeated all framework experiments ten times. The baseline method (BL) is to train a model on batch t with no resampling and use it to test on batch $t + 1$. For instructive purposes, we also examine the performance of positive class propagation when only the misclassified instances are propagated (SE-). We report several widely accepted imbalanced classification metrics including the area under the receiver operating characteristic curve (AUC-ROC) as implemented in WEKA [11], precision, recall, F_1-measure, and the area under the precision-recall curve (AUC-PR) as implemented in [12]. We observe that AUC-PR is a more discriminating measure than AUC-ROC in the domain of imbalance. Bold font indicates the highest value, but not necessarily statistical significance.

From Table 3(a) we see immediately that propagating misclassified negative class instances improves precision. The method creates a more complete boundary for training. In all cases, it is much closer to baseline precision, while sacrificing very little in recall. In some cases, it even improves precision beyond the baseline, which uses no resampling. Table 3(b) illustrates the trade-off performance on precision and recall and we see here that BD outperforms SE in terms of F_1-measure by at least 5 percent and as much as 26 percent. Finally, BD always performs at least as well as SE in AUC-ROC while outperforming it for every data set in the more discriminating AUC-PR measure.

For a broader view of the performance of BD with respect to the baseline and with respect to SE, we provide comprehensive coverage of 21 data sets from the UCI repository. To obtain these results, we simply execute each framework ten times on each data set. The data sets have precisely the properties described in 2. Each cell intersecting a method and a data set represents the rank of the method on that data set for the metric in the heading. A rank of 1 means the method performs best, and a rank of three means it performs worst. At the bottom, we provide the mean rank and the overall rank. The overall rank is calculated using the Nemenyi procedure as described in [13] with $\alpha = 0.05$. Statistical significance is indicated between two ranks when a value of one or greater separates them. This comes into play for the F_1-measure column where BD performs significantly better than BL, but SE does not perform statistically significantly better than BL, and BD does not perform statistically significantly better than SE.

Table 4. Performance on Static Data Sets

Name	Precision			Recall			F_1-measure			AUC-ROC			AUC-PR		
	BL	SE	BD	BL	SE	BD	BL	SE	BD	BL	SE	BD	BL	SE	BD
adult	1	3	2	3	1	2	2	3	1	3	2	1	3	2	1
boundary	3	2	1	3	1	2	3	2	1	3	2	1	3	2	1
breast-w	2	3	1	3	2	1	3	2	1	3	2	1	3	2	1
cam	1	2	3	3	1	2	3	1	2	3	1	2	3	1	2
covtype	2	3	1	3	1	2	3	2	1	3	2	1	3	2	1
estate	3	2	1	3	1	2	3	1	2	3	1	2	3	1	2
fourclass	3	2	1	3	1	2	3	2	1	3	2	1	3	2	1
german	1	2	3	3	1	2	3	1	2	3	1	2	3	2	1
ism	1	2	3	3	1	2	1	2	3	3	2	1	2	3	1
letter	1	3	2	3	2	1	1	3	2	3	2	1	3	2	1
oil	1	3	2	3	1	2	1	2	3	3	1	2	3	2	1
page	1	2	3	3	1	2	1	3	2	3	2	1	3	2	1
pendigits	1	3	2	3	1	2	2	3	1	3	2	1	3	2	1
phoneme	2	3	1	3	1	2	3	2	1	3	2	1	3	2	1
phoss	1	2	3	3	1	2	3	1	2	3	1	2	3	1	2
pima	1	3	2	3	1	2	3	1	2	3	2	1	3	2	1
satimage	1	3	2	3	1	2	3	2	1	3	2	1	3	2	1
segment	1	3	2	3	1	2	2	3	1	3	2	1	2	3	1
splice	3	2	1	3	2	1	3	2	1	3	2	1	3	2	1
svmguide1	1	3	2	3	1	2	3	1	2	3	1	2	3	1	2
thyroid	1	3	2	3	1	2	1	3	2	3	2	1	3	2	1
MEAN	1.5	2.6	1.9	3.0	1.1	1.9	2.4	2.0	1.6	3.0	1.7	1.3	2.9	1.9	1.2
OVERALL	1	2.5	1.5	3	1	2	2.5	2	1.5	2.5	2	1.5	2.5	2	1.5

Table 5. Performance on Drifting Data; Weighting Alone

(a) Precision and Recall

	Precision				Recall			
	SA	HD	HDIG	MW	SA	HD	HDIG	MW
adult	0.792	**0.793**	**0.793**	0.792	0.526	0.526	0.526	0.526
can	0.000	0.000	0.000	0.000	0.000	0.000	0.000	0.000
compustat	0.246	0.264	**0.345**	0.275	0.005	0.007	0.011	**0.018**
covtype	0.837	**0.838**	**0.838**	0.837	**0.683**	**0.683**	0.681	0.682
football	0.763	0.763	0.760	**0.764**	0.669	0.670	0.669	**0.671**
kddcup2008	0.038	0.046	0.046	**0.092**	0.005	0.013	0.013	**0.020**
ozone-1h	0.144	0.118	0.118	**0.150**	0.095	0.095	0.095	0.095
ozone-8h	0.154	**0.200**	0.200	0.155	0.061	**0.119**	**0.119**	0.061
STAGGER	0.656	0.656	0.656	**0.741**	0.454	0.454	0.454	**0.588**
text	0.657	**0.659**	**0.659**	0.641	0.693	0.693	0.693	**0.696**
wrds	0.971	0.971	0.971	0.971	**0.875**	**0.875**	0.861	**0.875**

(b) F_1-measure and PRBEP

	F_1-measure				PRBEP			
	SA	HD	HDIG	MW	SA	HD	HDIG	MW
adult	0.631	0.631	0.631	0.631	0.620	0.620	0.620	0.620
can	0.000	0.000	0.000	0.000	0.000	0.000	0.000	0.000
compustat	0.010	0.014	0.021	**0.034**	0.519	0.519	0.519	0.519
covtype	0.751	**0.752**	0.751	**0.752**	0.536	0.536	0.536	0.536
football	0.712	0.713	0.711	**0.714**	0.688	0.688	0.688	0.688
kddcup2008	0.009	0.017	0.017	**0.025**	0.160	0.160	0.160	0.160
ozone-1h	0.104	0.100	0.100	**0.106**	0.516	0.516	0.516	0.516
ozone-8h	0.072	**0.142**	**0.142**	0.072	0.534	0.534	0.534	0.534
STAGGER	0.508	0.508	0.508	**0.630**	0.711	0.711	0.711	0.711
text	0.671	**0.672**	**0.672**	0.664	0.534	0.534	0.534	0.534
wrds	**0.918**	**0.918**	0.904	**0.918**	0.462	0.462	0.462	0.462

(c) AUC-ROC and AUC-PR

	AUC-ROC				AUC-PR			
	SA	HD	HDIG	MW	SA	HD	HDIG	MW
adult	0.892	0.892	0.892	0.892	0.763	0.763	0.763	0.763
can	0.815	0.815	0.815	0.815	0.007	0.007	0.007	0.007
compustat	0.741	0.743	**0.744**	0.743	0.207	0.228	**0.251**	0.226
covtype	0.970	0.970	0.970	0.970	0.834	0.834	0.834	0.834
football	0.855	0.855	0.855	0.855	0.776	**0.777**	**0.777**	0.776
kddcup2008	0.780	**0.790**	0.780	0.767	**0.124**	0.122	0.122	0.121
ozone-1h	0.784	**0.795**	0.794	0.784	0.139	**0.158**	**0.158**	0.139
ozone-8h	**0.748**	0.747	0.746	0.744	0.175	**0.176**	0.175	0.174
STAGGER	0.744	0.739	0.739	**0.795**	0.807	0.807	0.807	**0.849**
text	**0.937**	**0.937**	**0.937**	0.934	0.617	**0.618**	**0.618**	0.610
wrds	0.973	0.973	0.973	0.973	0.698	0.698	0.698	0.698

We can conclude that both BL and BD achieve better precision than SE, which loses precision with respect to BL because it tends to push the border beyond more negative class instances. Although BL usually achieves higher precision than BD, the results are not significant with $\alpha = 0.05$. Although we see that SE outranks BD in terms of recall, a quick look at both Table 3(a) and much of the data underlying the ranks show that the benefit of SE over BD with respect to BL is often negligible given the precision gains. Although individual F_1-measure results suggest that BD often outperforms SE, we cannot say this with $\alpha = 0.05$, nor can we say with confidence that SE outperforms BL. Finally, on both AUC-ROC and AUC-PR, despite frequent wins by BD, the results are not statistically significant at $\alpha = 0.05$. For AUC-PR, at $\alpha = 0.10$ we can say that SE performs worse than BD.

Next, we examine how these methods perform on data that exhibits distributional or concept drift, especially in the context of class imbalance. As control factors, we include adult and kddcup2008, which exhibit no form of data drift. We also include the benchmark data set, STAGGER, which exhibits only concept drift and no distributional drift. STAGGER also does not exhibit meaningful imbalance. Table 5 illustrates results for simple average voting (SA), ensembles weighted using Hellinger distance (HD), ensembles weighted using the HDIG distance (HDIG), and ensembles weighted using a meta-classification scheme (MW).

These results indicate that applying weights to ensemble members instead of using the simple average of unweighted probability outputs without any underlying sampling does little to improve classification. The authors of [14] found a similar result in which modifications to a majority voting scheme produced small and inconsistent improvements. compustat is an exemplary data set for showing the usefulness of the HD-IG method. Many of the performance metrics show large differences. The exemplary data set for exhibiting the usefulness of the MW method is STAGGER. Its posterior drift is detected by the performance features in the performance history data.

Applying ensemble weights seems to produce much more consistent results after resampling. This result is surprising since resampling changes the distribution of the models without changing the distribution of the underlying data used for the distance computation. It is meaningful because HD-IG systematically improves the recall and F_1-measure for almost all of the data sets, although the improvements are minor and apply almost equally to the randomized data in adult and covtype.

10   Timing and Parallelism

Because stream mining frameworks and algorithms are often employed in time-critical or resource constrained situations, it is more important than in typical non-streaming applications that they perform efficiently. Aside from any specific performance or resource constraints, mining data streams has an inherent requirement that the classification task have a throughput at least equal to the rate at which instances or batches arrive. Research like that in [15] even explores the topic specifically.

Table 6. Performance on Drifting Data; Weighting and Resampling

(a) Precision and Recall

	Precision			Recall		
	BD	BD-HDIG	BD-MW	BD	BD-HDIG	BD-MW
adult	0.614	**0.618**	0.614	0.781	**0.783**	0.776
can	0.027	0.027	0.027	0.011	**0.014**	0.012
compustat	0.142	**0.161**	0.128	0.326	**0.396**	0.374
covtype	0.728	**0.734**	0.728	0.892	**0.899**	0.897
football	0.774	0.774	0.774	0.651	**0.655**	0.650
kddcup2008	**0.071**	0.062	0.070	0.370	**0.379**	**0.379**
ozone-1h	0.063	**0.074**	0.055	0.455	**0.552**	0.431
ozone-8h	0.133	0.134	**0.149**	0.422	**0.428**	0.425
STAGGER	0.569	0.598	**0.686**	0.454	0.454	**0.588**
text	0.510	0.511	**0.558**	**0.831**	0.816	0.821
wrds	0.973	0.973	**0.974**	**0.874**	0.801	0.873

(b) F_1-measure and PRBEP

	F_1-measure			PRBEP		
	BD	BD-HDIG	BD-MW	BD	BD-HDIG	BD-MW
adult	0.686	**0.689**	0.685	0.783	**0.785**	0.784
can	0.012	**0.016**	0.014	0.021	0.021	0.021
compustat	0.185	**0.217**	0.183	0.701	0.701	0.701
covtype	0.796	**0.802**	0.798	0.708	0.709	0.709
football	0.707	**0.709**	0.706	**0.782**	0.781	**0.782**
kddcup2008	**0.105**	0.098	**0.105**	**0.325**	0.323	0.323
ozone-1h	0.110	**0.129**	0.096	0.701	0.701	0.701
ozone-8h	0.199	0.203	**0.218**	0.700	0.700	0.700
STAGGER	0.475	0.488	**0.607**	0.815	**0.818**	0.813
text	**0.629**	0.611	0.625	0.700	0.700	0.700
wrds	**0.918**	0.854	**0.918**	0.918	0.918	**0.919**

(c) AUC-ROC and AUC-PR

	AUC-ROC			AUC-PR		
	BD	BD-HDIG	BD-MW	BD	BD-HDIG	BD-MW
adult	0.905	**0.906**	0.905	0.783	**0.784**	0.783
can	0.819	**0.821**	0.819	0.026	0.026	0.026
compustat	0.770	0.774	**0.780**	0.207	**0.210**	0.184
covtype	**0.982**	0.981	**0.982**	0.866	0.866	0.866
football	**0.862**	0.861	**0.862**	**0.795**	0.794	**0.795**
kddcup2008	0.931	**0.940**	0.936	0.201	**0.225**	0.196
ozone-1h	0.751	**0.785**	0.727	**0.108**	0.102	0.095
ozone-8h	0.724	**0.728**	0.725	0.159	0.165	**0.168**
STAGGER	0.709	0.704	**0.753**	0.788	0.785	**0.812**
text	0.941	0.936	**0.942**	**0.625**	0.617	0.622
wrds	**0.974**	0.973	**0.974**	**0.975**	0.974	**0.975**

The increased time requirement for BD is minimal but evident since it must consider the misclassified negative class instances through time. We demonstrate here that the performance benefit of DW is offset by a modest increase in computational requirements. We begin with a theoretical examination and move to an empirical study. Hellinger distance may be computed on a single feature in $O(e)$ where e is the number of instances in the batch. Computing Hellinger distance for all features is $O(|\boldsymbol{f}| \cdot e)$. As each batch with unclassified instances arrives, all previous batches must be compared with only the new batch. The number of Hellinger distance computations for each batch therefore increases linearly with the number of batches. For a batch i there will be $O(|\boldsymbol{f}| \cdot e) \cdot O(i)$ operations for ensemble construction. Because the information gain component of the distance measure can be saved for each feature in each batch after computation, its growth through time is described by $O(1)$ and its time requirement for each batch is $O(e)$, which is subsumed by the Hellinger distance computation. The overhead requirement for preparing the weights for all batches in a stream up to and including batch n is such:

$$\sum_{i=0}^{n} O(|\boldsymbol{f}| \cdot e + 1) = \frac{n \cdot (n+1)}{2} \cdot O(|\boldsymbol{f}| \cdot e) = O\left(n^2 \cdot |\boldsymbol{f}| \cdot e\right) \qquad (5)$$

The task also enjoys simple parallelization. Distance computations for a batch b_i and the incoming test batch b_{t+1} are entirely independent and may easily be done in parallel. Each of the ensemble members can run independently to produce its probability estimations or classification output on new unlabeled instances.

We include BL, SE, BD, and HD for comparison. We omit HD-IG because information gain computation results can be saved resulting in negligible contributions to time requirements. We also omit MW since its complexity is dependent on the choice of classifier and the specific nature of the problem at hand, although we do observe that its growth will be at least linear growth in the number of batches. Theoretically, BL should exhibit no increase in time requirements with incoming batches. SE should exhibit small increases with each batch as SE requires all previously constructed models to provide a probability estimation and training time increases for the increasingly large resampled batches.

For empirical study, we use kddcup2008 data only because the data volume is sufficient to accommodate 80 batches of 1278 instances each. Times include the entire training and evaluation process for each batch. We made no effort to improve the performance of any of the algorithms over any others. We observe that all methods except BL theoretically exhibit, with varying coefficients, linear growth in the number of batches over time. Finally, Figure 6 shows all methods requiring more time than BL for almost all batches. This is only a consequence of the fact that we divided the data into batches with small numbers of instances so that sufficient batches would exist to illustrate the growth patterns. In reality, the BL method will require longer whenever methods that undersample remove a significant portion of the batch.

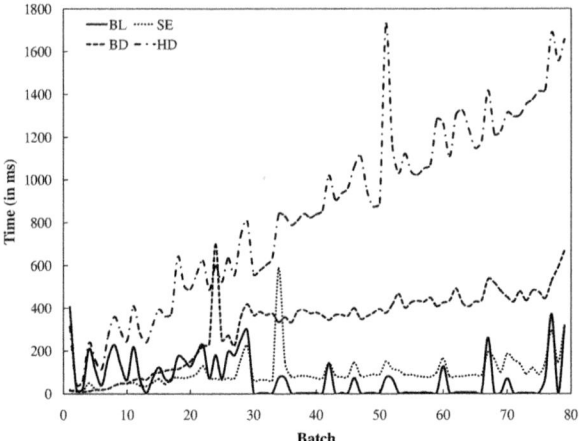

Fig. 6. The time requirement in milliseconds for processing batches from 1 through 80

11   Conclusions

Generating classifier ensembles to detect and address concept drift in a proactive manner is a powerful new solution. It performs dramatically better than existing methods on extremely complicated data streams with complex concept drift and high degrees of imbalance. Simultaneously, it sacrifices no significant performance on simpler data sets with no concept drift and only moderate imbalance. Often classification improvements are measured in a few percentage points of accuracy or small fractional increases in AUC-ROC. Our method achieves gains in several performance metrics that are as great as an order of magnitude with respect to preexisting methods of handling the intersecting problems of concept drift and imbalance.

We acknowledge that the assumption that feature relevance is stable over time does not always hold. Several works in the field observe that it is not [16,17,18,19]. We can say that for all the data sets on which we tested, the information gain adjusted distance function performed better than Hellinger distance alone. For arbitrarily complex drifts in $\Delta P(f)$ or drifts in $\Delta P(c|f)$ signaled by $\Delta P(f)$, distributional divergence performs well. It is theoretically impossible to detect with certainty that a new batch contains a posterior probability shift. In situations like these, it may become necessary to sample a few instances from the upcoming batch as in [20]. Another alternative is to combine performance-based weighting with the weighting proposed here.

We believe that the distance measure proposed herein is an excellent way to measure the distance between two data sets. In its current form, it only uses information gain from one of the two data sets to provide a distance, but it can easily be extended to accept information gain from two sources when class labels are available from both. We hope to compare this measure of data set distance to several others to compare properties and usefulness.

12   Related Work

Researchers have extensively studied class imbalance. The most popular approaches for handling imbalance are various forms of resampling to achieve a more balanced distribution, including random undersampling of the negative class and SMOTE [21]. A more recent approach is an active learning system capable of efficiently querying large data sets to find informative examples [22].

Together, [23] and [24] serve as an excellent overview of the principles of concept drift. An early seminal paper in overcoming concept drift was [25]. Since then, there has been much research in data stream mining to attempt to react to concept drift as quickly as possible. Many batch approaches employ some form of ensemble technique. Some use simple average [5], while others employ weighting [3,6]. With various levels of sophistication, some systems maintain a sliding window of examples or models. [26] and [27] dynamically adjust window size and the former provides strict performance guarantees. We observe that a contiguous window, even an optimally sized one, may fail to contain recurring concepts if they were previously forgotten by the system. This concept of periodicity is briefly discussed in [1]. The FLORA systems have the ability to reactivate old concepts outside the window [4].

We are aware of few publications that directly address the notion of class imbalance in combination with concept drift in data streams. The FLORA systems include distinct rules for positive examples [4], but do not directly target them. Work in [28] focuses on detecting concept drift in potentially adversarial scenarios such as the perpetration of fraud. In [5], which specifically addresses imbalance, the primary contribution was the idea of propagating positive class examples to overcome class imbalance.

Much of our work focuses on distributional divergence. Although the topic has not been explored extensively in general stream mining, it is a familiar concept in the related field of novelty detection, which is the task of realizing the occurrence of previously unobserved or infrequently observed concepts. An excellent review of distributional measures in novelty detection appears in [29]. Although it does not implement or report results for any methods, [24] addresses the potential of using distributional considerations to detect concept drift. The work in [30] proposes modeling the underlying distributions in data and using the models to detect change. In [31], distributional measures are used for ensemble weighting. Finally, [20] uses distributional measures in the forms of expected loss due to feature dissimilarity and decision tree leaf statistics to determine when data distributions are changing outside an acceptable level.

References

1. Becker, H., Arias, M.: Real-time ranking with concept drift using expert advice. In: Proceedings of the 13th ACM SIGKDD international conference on Knowledge discovery and data mining, pp. 86–94. ACM Press, New York (2007)
2. Hulten, G., Spencer, L., Domingos, P.: Mining time-changing data streams. In: KDD 2001: Proceedings of the seventh ACM SIGKDD international conference on Knowledge discovery and data mining, pp. 97–106. ACM, New York (2001)

3. Kolter, J.Z., Maloof, M.A.: Dynamic weighted majority: An ensemble method for drifting concepts. Journal of Machine Learning Research 8, 2755–2790 (2007)
4. Widmer, G., Kubat, M.: Learning in the presence of concept drift and hidden contexts. Machine Learning 23, 69–101 (1996)
5. Gao, J., Fan, W., Han, J., Yu, P.S.: A general framework for mining concept-drifting data streams with skewed distributions. In: SDM 2007: Proceedings of the SIAM International Conference on Data Mining (2007)
6. Wang, H., Fan, W., Yu, P.S., Han, J.: Mining concept-drifting data streams using ensemble classifiers. In: KDD 2003: Proceedings of the ninth ACM SIGKDD international conference on Knowledge discovery and data mining, pp. 226–235. ACM, New York (2003)
7. Han, H., Wang, W.Y., Mao, B.H.: Borderline-smote: A new over-sampling method in imbalanced data sets learning. Advances in Intelligent Computing, 878–887 (2005)
8. Cieslak, D.A., Chawla, N.V.: Detecting fractures in classifier performance. In: ICDM 2007: Seventh IEEE International Conference on Data Mining, pp. 123–132 (2007)
9. Cieslak, D.A., Chawla, N.V.: Learning decision trees for unbalanced data. In: European Conference on Machine Learning. Springer, Heidelberg (2008)
10. Asuncion, A., Newman, D.: Uci machine learning repository (2007)
11. Witten, I.H., Frank, E.: Data Mining: Practical Machine Learning Tools and Techniques, 2nd edn. Morgan Kaufmann, San Francisco (2005)
12. Davis, J., Goadrich, M.: The relationship between precision-recall and roc curves. In: ICML 2006: Proceedings of the 23rd international conference on Machine learning, pp. 233–240. ACM, New York (2006)
13. Demšar, J.: Statistical comparisons of classifiers over multiple data sets. The Journal of Machine Learning Research 7, 1–30 (2006)
14. Street, N.W., Kim, Y.: A streaming ensemble algorithm (sea) for large-scale classification. In: KDD 2001: Proceedings of the seventh ACM SIGKDD international conference on Knowledge discovery and data mining, pp. 377–382. ACM, New York (2001)
15. Haghighi, P.D., Gaber, M.M., Krishnaswamy, S., Zaslavsky, A., Seng, L.: An architecture for context-aware adaptive data stream mining. In: Kok, J.N., Koronacki, J., Lopez de Mantaras, R., Matwin, S., Mladenič, D., Skowron, A. (eds.) ECML 2007. LNCS (LNAI), vol. 4701. Springer, Heidelberg (2007)
16. Blum, A.: Empirical support for winnow and weighted-majority algorithms: Results on a calendar scheduling domain. Machine Learning 26, 5–23 (1997)
17. Forman, G.: Tackling concept drift by temporal inductive transfer. In: SIGIR 2006: Proceedings of the 29th annual international ACM SIGIR conference on Research and development in information retrieval, pp. 252–259. ACM, New York (2006)
18. Harries, M., Horn, K.: Detecting concept drift in financial time series prediction using symbolic machine learning. In: Eighth Australian Joint Conference on Artificial Intelligence, pp. 91–98. World Scientific Publishing, Singapore (1995)
19. Widmer, G.: Tracking context changes through meta-learning. Machine Learning 27, 259–286 (1997)
20. Fan, W., Huang, Y.a., Wang, H., Yu, P.S.: Active mining of data streams. In: Proceedings of the Fourth SIAM International Conference on Data Mining, Society for Industrial Mathematics, pp. 457–461 (2004)
21. Chawla, N.V., Bowyer, K.W., Hall, L.O., Kegelmeyer, P.W.: Smote: Synthetic minority over-sampling technique. Journal of Artificial Intelligence Research 16, 341–378 (2002)

22. Ertekin, S., Huang, J., Bottou, L., Giles, L.: Learning on the border: Active learning in imbalanced data classification. In: CIKM 2007: Proceedings of the sixteenth ACM Conference on information and knowledge management, pp. 127–136. ACM, New York (2007)
23. Kelly, M.G., Hand, D.J., Adams, N.M.: The impact of changing populations on classifier performance. In: KDD 1999: Proceedings of the fifth ACM SIGKDD international conference on Knowledge discovery and data mining, pp. 367–371. ACM, New York (1999)
24. Kuncheva, L.I.: Classifier ensembles for detecting concept change in streaming data: Overview and perspectives. In: Proceedings of the 2nd Workshop SUEMA 2008 (ECAI 2008), pp. 5–10 (2008)
25. Schlimmer, J.C., Granger, R.H.: Incremental learning from noisy data. Machine Learning 1, 317–354 (1986)
26. Bifet, A., Gavaldá, R.: Learning from time-changing data with adaptive windowing. In: SIAM International Conference on Data Mining, SDM 2007 (2006)
27. Klinkenberg, R.: Using labeled and unlabeled data to learn drifting concepts. In: Workshop notes of the IJCAI 2001 Workshop on Learning from Temporal and Spatial Data, pp. 16–24 (2001)
28. Phua, C., Miles, K.S., Lee, V., Gayler, R.: Adaptive spike detection for resilient data stream mining. In: Proceedings of the sixth Australasian conference on Data mining and analytics (AusDM 2007), pp. 181–188. Australian Computer Society, Inc., Darlinghurst (2007)
29. Markou, M., Singh, S.: Novelty detection: A review - part 1: Statistical approaches. Signal Processing 83, 2481–2497 (2003)
30. Korn, F., Muthukrishnan, S., Wu, Y.: Modeling skew in data streams. In: SIGMOD 2006: Proceedings of the 2006 ACM SIGMOD international conference on Management of data, pp. 181–192. ACM, New York (2006)
31. Nishida, K., Yamauchi, K., Omori, T.: Ace: Adaptive classifiers-ensemble system for concept-drifting environments. Multiple Classifier Systems, 176–185 (2005)

Two Measures of Objective Novelty in Association Rule Mining*

José L. Balcázar

Departamento de Matemáticas, Estadística y Computación
Universidad de Cantabria
Santander, Spain
joseluis.balcazar@unican.es

Abstract. Association rule mining is well-known to depend heavily on
a support threshold parameter, and on one or more thresholds for intensity of implication; among these measures, confidence is most often
used and, sometimes, related alternatives such as lift, leverage, improvement, or all-confidence are employed, either separately or jointly with
confidence. We remain within the support-and-confidence framework in
an attempt at studying complementary notions, which have the goal of
measuring relative forms of objective novelty or surprisingness of each
individual rule with respect to other rules that hold in the same dataset.
We measure novelty through the extent to which the confidence value is
robust, taken relative to the confidences of related (for instance, logically
stronger) rules, as opposed to the absolute consideration of the single rule
at hand. We consider two variants of this idea and analyze their logical
and algorithmic properties. Since this approach has the drawback of requiring further parameters, we also propose a framework in which the
user sets a single parameter, of quite clear intuitive semantics, from which
the corresponding thresholds for confidence and novelty are computed.

1 Introduction and Related Work

Association rule mining is a process by which a transactional or relational dataset
is explored in an attempt at identifying implications among its elementary components (items or attribute values). The syntax of implications is very sugges-
tive of cause-effect relationships; therefore, such syntax is welcome by human
decision-makers and domain experts, who can analyze actions to be taken on
the basis of the causality intuitively suggested by the implications found.

The idea of expressing knowledge extracted from data in a form of implications has been proposed in a myriad of contributions, many of these in a manner
independent of each other. An early development, largely unknown, that already
offered the current notion of association rules as a mere part of a much more
expressive logic-based system is described in [20]. The research area of Machine

---

* This work has been partially supported by project TIN2007-66523 of Programa
Nacional de Investigación (FORMALISM), Ministerio de Ciencia e Innovación
(MICINN), Spain, and by the Pascal-2 Network of the European Union.

T. Theeramunkong et al. (Eds.): PAKDD Workshops 2009, LNAI 5669, pp. 76–98, 2010.
© Springer-Verlag Berlin Heidelberg 2010

Learning has contributed also many algorithms to "learn rules from examples", which, often, amounts to identifying implications or variants thereof. Purely logical implications have been explored in many contributions (see [16], [35], and the references there for one of the settings, and [23], [25] for closely related perspectives); a proposal that gave the topic of research and applications of association rules inmense momentum, was the description in [2] of the usefulness of parameterizing the association mining process according to a support constraint and a confidence constraint (or "precision" in [30]). In fact, the support constraint opened the door to the design of practically feasible algorithms, starting with [3]; in fact, different datasets often require different algorithmics: see the outcomes of the FIMI competition [14] and the alternatives described in the survey [12]. On the other hand, there is a clear need of quantifying "degrees of implication" because purely logical implications turn out not to match exactly the needs of practical association mining projects. However, several criticisms could be put forward about confidence as a measure of "degree of implication", and a large number of alternatives have been proposed, evaluated, and studied; the literature about these notions is, in fact quite large [17], [19], [21], [37]. A good survey with many references is [18].

Yet, we prefer to develop our proposal in the context of support and confidence bounds, for several reasons. First, conditional probability is a concept known to many educated users from a number of scientific and engineering disciplines, so that communication with the data mining expert is simplified if our measure is confidence. Second, as a very elementary concept, it is the best playground to study other proposals, such as our contribution here, which could be then lifted to other similar parameters. Third, we believe that, in fact, our relative measures will make up for many of the objections raised against confidence. Additionally, it must be taken into account that the quantity of data is usually insufficient to test the extremely large number of hypotheses given by the set of all possible rules, even if schemes more efficient than the Bonferroni guarantees are employed; and it has been observed and argued that the combination of support and confidence is already very good at discarding rules that are present only as statistical artifacts and do not really correspond to correlations in the phenomenon at the origin of the dataset [32].

Now, let us put forward the following considerations. The outcome of a data mining project is expected to offer some degree of novelty. A wide spectrum of subjective considerations regarding the user's previous knowledge can be considered, and, of course, novelty with respect to knowledge existing previously to the data mining process is hard to formalize. But one fact is clear: novelty cannot be evaluated in an absolute form; it refers to knowledge that is somehow unexpected, and therefore some expectation, lower than actually found, must exist, due to some alternative prediction mechanism. Additionally, an intuitive "rule of thumb" is that the amount of novel facts must be low in order that novelty is actually useful.

We propose to measure the novelty of each rule with respect to the rest of the outcome of the same data mining process. To do this, we resort to recent advances in the construction of irredundant bases and in mathematical characterizations of the most natural notion of redundancy. As we shall see, a redundant rule is so because we can know beforehand, from the information in a basis, that its confidence will be above the threshold. Pushing this intuition further, an irredundant rule in the basis is so because its confidence is higher than what the rest of the basis would suggest: this opens the door to asking, "how much higher?". If the basis suggests, say, a confidence of 0.8 (or 80%) for a rule, and the rule has actually a confidence of 0.81, the rule is indeed irredundant and brings in additional information, but its novelty, with respect to the rest of the basis, is not high; whereas, in case its confidence is actually 0.95, quite higher than the 0.8 expected, the fact can be considered novel, in that it states something really different from the rest of the information mined. We provide a new notion that formalizes this intuition, and show that it indeed refines very much the data mining process, but has a limitation due to being too close to a fully logical approach. Then we relax slightly the definition into a more useful variant, and we study both concepts.

The main notions to be defined below are similar to the "pruning" proposal from [29], in that the intuition is the same; two major differences are, first, that we will work on an already heavily reduced basis, so that a large portion of the pruning becomes unnecessary, and that for what remains, the pruning in [29] is based on the χ^2 statistic, whereas we will look instead into the confidence thresholds that would make the rule logically redundant. Our notions are also similar to the notion of *improvement*, proposed in [7] (and also briefly discussed in [29], although we are not aware of that proposal having received further attention); this quantity also attempts at discarding uninteresting rules due to the same intuitions as ours; but it is a measure of an absolute, additive confidence increase, with no reference to representative rules or standard redundancy, and it only allows for varying the antecedent into a smaller one, keeping the same consequent. Our quotient-based definitions are more powerful, enjoy better algorithmic properties than those currently known for the analogous difference-based alternative, and are also, in our opinion, more natural.

Our notions have some surface similarity as well with the notion of all-confidence [33] and the related concept of m-patterns [31]. However, these notions are rather restrictive, and provide only strong "niches" where all the sets of attributes within an output pattern depend heavily pairwise among them. We wish to depart in a lesser degree from the standard association rule setting. On the other hand, a strong point of these notions is that they bring in an antimonotonicity property to prune the search space. Instead, we just employ a support bound for its antimonotonicity property, and discuss our contribution in terms of postprocessing the output of a standard frequent closed set miner.

Each of these additional measures, and ours are not exceptions, raises an additional difficulty. For a vast majority of datasets, already the setting of a support and confidence value by a human requires enormous expertise and intuition, and/or insistingly repeated runs of the computation process with different values. Few works discuss the setting of the support threshold for association rules; worth mentioning are the works [6], [13], [22], and [24], all of which provide interesting advances for the case where the association rules are to be used as a classifier (which is not our case here), using the additional information that one of the attributes will be a as target class; this opens the door to using coverage analysis or criteria related to the ROC curve to orient the decision of which support threshold to use. Many algorithms related to Machine Learning have a similar criticism; say, the parameter corresponding to the box constraint of the soft-margin support vector machines, as one mere example. Many successful algorithms are so through the identification of some sort of autonomic or semi-autonomic self-adjustment of the parameters, thus freeing the user from having to choose a value for them.

However, fully removing all parameters would not be the best choice either. It is clear that different characteristics of datasets (largish or smallish transactions, large or small deviations from the average in the transaction sizes, large or small universe of items, more uniform or less uniform distributions of the individual items) are likely to call for somewhat tailored explorations. Therefore, the data mining process needs some way of tuning the exploration to the dataset at hand. We propose here an interpretation of confidence that allows us to suggest values for the bounds on our new novelty parameters, automatically from the confidence bound.

1.1 Redundancy among Association Rules

We start our analysis from one of the notions of redundancy defined formally first in [1], but employed also, generally with no formal definition, in several papers on association rules; thus, we will qualify sometimes this redundancy notion as "standard". We give up front two equivalent characterizations of the notion: the second one was proposed, as nearly identical "covering"-like simplifications, in several independent sources ([1], [26], [36]); the fact that they are equivalent to standard redundancy, instead of being a simplified variant of it, is quite recent [5].

We denote itemsets by capital letters from the end of the alphabet, and use juxtaposition to denote union, as in XY. For a given dataset $\mathcal{D}$, consisting of transactions, each of which is an itemset labeled with a unique transaction identifier, we can count the *support* $s(X)$ of an itemset X, which is the cardinality of the set of transactions that contain X. The *confidence* of a rule $X \rightarrow Y$ is $c(X \rightarrow Y) = s(XY)/s(X)$.

Lemma 1. *[5] Consider two association rules, $X_0 \rightarrow Y_0$ and $X_1 \rightarrow Y_1$. The following are equivalent:*

1. *The confidence and support of $X_0 \rightarrow Y_0$ are always larger than or equal to those of $X_1 \rightarrow Y_1$, in all datasets; that is, for every dataset $\mathcal{D}$, $c(X_0 \rightarrow Y_0) \geq c(X_1 \rightarrow Y_1)$ and $s(X_0Y_0) \geq s(X_1Y_1)$.*
2. $X_1 \subseteq X_0 \subseteq X_0Y_0 \subseteq X_1Y_1$.

The fact that 2 implies 1 is easy to see and was pointed out in the references indicated. The fact that 1 implies 2 is nontrivial and much more recently shown. Whenever rules $X_0 \rightarrow Y_0$ and $X_1 \rightarrow Y_1$ fulfill either of the two equivalent conditions, we say that $X_0 \to Y_0$ is *redundant* with respect to $X_1 \to Y_1$. As an example, for items A, B, C, and D, the rule $AB \to C$ is redundant with respect to rule $A \to BC$, and is also redundant with respect to $AB \to CD$. The first of the two equivalent forms of definition is akin to the definition of entailment in purely logic-based studies, and we will use sometimes the phrase "logically stronger" to refer to a rule that makes another one redundant with respect to standard redundancy.

Note that the rules $X \to Y$ and $X \to XY$ are mutually redundant, in fact fully equivalent because their confidence $s(XY)/s(X)$ and support $s(XY)$ always coincide. Therefore we consider all association rules where *the right-hand side always includes the left-hand side*, although for the purpose of showing them to the user the repeated items of the left-hand side will be removed from the right-hand side. This simple convention greatly simplifies the mathematical development.

There are several alternative notions of redundancy in the literature; see [5] for further comparisons among a few of them. For this particular notion we have just given, the aim is clear: whatever the dataset under analysis, and the support and confidence parameters, if we find that rule $X_1 \to Y_1$ appears among the mined rules by passing the support and confidence thresholds, any other rule $X_0 \to Y_0$ showing standard redundancy with respect to it is known to be also in the set of mined rules without need to inspect them to check out. This is because the support and confidence must be at least the same as those of rule $X_1 \to Y_1$, whence it passes the thresholds as well.

1.2   Representative Rules

The fact that the output of association rule miners tends to be far larger than desired has been widely reported; it is also self-apparent to anyone that has tried any of the association miners in data mining packages or implementations freely available on the web, e.g. [9].

Our implementation builds on the *representative rules* for association rules, proposed independently and in different but equivalent ways, in [1], in [26], and in [36]. Recently, several new mathematical properties of this basis have been proved, including a form of optimality [5].

Definition 1. *Fix a dataset $\mathcal{D}$ and confidence and support thresholds. The corresponding basis of* representative rules *consists of all the rules that hold in $\mathcal{D}$, passing both thresholds, which are* not *redundant with respect to any other rule that holds in $\mathcal{D}$ for the same thresholds.*

Among several equivalent possibilities to define representative rules, we have chosen a definition so that the following claim becomes intuitively clear: every rule that passes the thresholds for $\mathcal{D}$ is either a representative rule, or is redundant with respect to a representative rule. Indeed, any given rule that is not among the representative rules must be redundant with respect to some other rule, which again must be redundant with respect to a third, and so on, until finiteness enforces termination that can be only reached by finding a rule in the basis, making redundant all the others found along the way. The formalization of this argument can be found in [26] (Lemma 1 must be taken into account to complete the proof).

Thus, every rule that passes the thresholds for $\mathcal{D}$ is either a representative rule, or is redundant with respect to a representative rule. Moreover, any basis, that is, any set of rules that makes redundant all the rules mined from $\mathcal{D}$ at the given thresholds, must include all the representative rules, since there is no other way of making them redundant. Thus, the representative rules form the unique smallest basis with respect to standard redundancy. (This is not true of rules of confidence 100%; for these absolute implications, the representative basis from [1], [26], [36] can be constructed as well and coincides with the "canonical iteration-free basis" of [38], the nonredundant implications of [40], the proposal in [35] and the "generic" (or "exact min-max") basis of [34]; but all these equivalent proposals fail to reach a minimum size, since there is a more economical alternative [15]. Full discussion can be found in [5], where all these facts, and also the equivalence of our formulation with the original ones, are studied in detail.)

In a sense, representative rules are sort of a required starting point, since they give demonstrably the best basis size one can hope for with no loss of information, with respect to redundancy as defined. Representative rules turn out to be intimately related to closed itemsets and minimal generators. These two notions play an important role in rule mining ([11], [27], [34], [40], [41]). A set is *closed* if there is no proper superset with the same support. A set is a *minimal generator* (or also a *free set*) if there is no proper subset with the same support. In the presence of a support threshold, frequent closed sets are closed sets whose support clears the threshold. Frequent closed sets are very crucial to the algorithmics of association rules and to the identification of irredundant bases. Specifically, in [27] we find a proof of the following nonobvious fact: all representative rules have a minimal generator as antecedent and a closed itemset as consequent (however, not all such pairs give representative rules). Good algorithms and implementations to find them already exist. Absolute optimality of certain versions of these bases is shown in [5].

2   Confidence Width

This section describes the foundations of our proposal. Our intuition is as follows: consider a rule $X \to Y$ of a given confidence, say $c(X \to Y) = c_0 \in [0, 1]$, in a given dataset $\mathcal{D}$. Assume that a fixed support threshold is enforced throughout the discussion, and consider what happens as we vary the confidence threshold γ.

If we set it higher than c_0, that is, $c_0 < \gamma$, the rule at hand will not play any role at all, being of confidence too low for the threshold. As we lower the threshold and reach exactly $\gamma = c_0$, the rule becomes part of the output of any standard association mining process, but two different things may happen: the question is whether, at the same confidence, some other "logically stronger" rule appears. If not, $X \to Y$ will belong to the representative rules basis for that threshold; but it may be that, at the same threshold, some other logically stronger rule is found. For instance, it could be that both rules $A \to B$ and

$A \to BC$ have confidence c_0: then $A \to B$ is redundant and will not belong to the basis for that confidence.

Let's then assume that the rule at hand does appear among the representative rules at the confidence threshold given by its own confidence value; and let's keep decreasing the threshold. At some lower confidence, a logically stronger rule may appear. If a logically stronger rule shows up early, at a confidence threshold γ very close to c_0, then the rule $X \to Y$ is not very novel: it is too similar to the logically stronger one, and this shows in the fact that the interval of confidence thresholds where it is a representative rule is short.

To the contrary, a stronger rule may take long to appear: in that case, only rules of much lower confidence entail $X \to Y$, so the fact that it does reach confidence c_0 is novel in this sense. The interval of confidence thresholds where $X \to Y$ is a representative rule is large. For instance, if the confidence of $A \to AB$ is 0.9, and all rules that make it redundant all have confidences below 0.75, the rule is a much better candidate to novelty than it would be if some rule like $A \to ABC$ would have a confidence of 0.88.

This motivates the following definition:

Definition 2. *Fix a dataset $\mathcal{D}$ and a support threshold τ. Consider a rule that has support at least τ in $\mathcal{D}$, say rule $X \to Y$. Consider all rules that are not equivalent to $X \to Y$, but such that $X \to Y$ is redundant with respect to them, and pick one with maximum confidence in $\mathcal{D}$ among them, say $X' \to Y'$ (thus $c(X' \to Y') \leq c(X \to Y)$). The* confidence width *of $X \to Y$ in $\mathcal{D}$ is:*

$$w(X \to Y) = \frac{c(X \to Y)}{c(X' \to Y')}$$

In case $X \to Y$ is representative, only rules of confidence smaller than γ can make it redundant. In order to check for the existence of $X' \to Y'$, one should mine at lower confidence levels (but see comments after Theorem 1 below). The confidence width can be defined equivalently as the ratio between the extremes of the interval of confidence thresholds that allow the rule to be representative. That is: the highest value where the rule can belong to the representative rule basis is the confidence of the rule; and the denominator is the highest value where there is a different representative rule that makes it redundant, thus forcing it out of the representative basis.

Observe that when $X \to Y$ is redundant with respect to $X' \to Y'$, its confidence must be at least the confidence of the latter, which implies that the confidence width is always greater than or equal to 1. For a rule $X'' \to Y''$, the confidence width is exactly 1 if and only if there is a rule making redundant $X'' \to Y''$ and having the same confidence: this is the same as saying that $X'' \to Y''$ is never among the representative rules. Regarding upper bounds, in principle there is none, as it may happen that a rule of as large confidence as desired is only redundant with respect to rules of as low confidence as desired.

2.1 Properties and Algorithms

We proceed to study some properties of the confidence width; by combining them with known properties of the standard redundancy and of the representative rules, we will obtain reasonably efficient ways to compute the width of the rules in the basis. We will need a preliminary fact:

Proposition 1. *Consider a rule $X \to Y$ and a different rule $X' \to Y'$ that makes it redundant; assume $X' \to Y'$ has maximum confidence as in the definition of width, say δ. Then $X' \to Y'$ can be chosen among the representative rules for confidence δ.*

This proposition can be proved easily by resorting to the known fact [26] that every rule of confidence δ is redundant with respect to a representative rule of the same confidence (possibly itself). As indicated in the previous section, rules not in the representative basis have minimum width, namely 1. Thus, to know the confidence width of all the rules it suffices to find it for representative rules.

We do not need to scan all frequent sets, since, as indicated above, it is known that if $X \to Y$ is a representative rule, then XY is a closed set and X is a minimal generator [27]. There are several published algorithms that compute the frequent closed sets and the minimal generators (see the survey [12]); in one form or another, all of them employ the key and well-known fact of the antimonotonicity of the frequent itemsets. These closures and minimal generators can be used to find the representative rules whose width is to be computed, by using the algorithm in [27].

A naive algorithm follows immediately: construct the representative rules and scan them repeatedly, applying Proposition 1 to find, for each rule $X \to Y$, the largest confidence c of any representative rule that makes $X \to Y$ redundant; use Lemma 1 to test for standard redundancy. Once this largest confidence c is known, the width of $X \to Y$ is clearly $w(X \to Y) = \frac{c(X \to Y)}{c}$ by definition. However, notice that this algorithm requires time quadratic in the number of representative rules, and that we mean *all* representative rules, that is, for *all* values of the confidence threshold. This is likely to be a large set.

2.2 An Alternative Algorithm

In some cases, we are likely to wish a computational shortcut: consider the usual case of a user having indicated thresholds for support and confidence, so that our proposal would end in answering the user with a set of representative rules that pass both thresholds, maybe ordered according to width, or possibly even pruned once more at a width threshold. In principle, we only need representative rules at the confidence threshold given. However, to compute the width, we need all representative rules at all threshold levels. If the threshold is somewhat high, say 0.8, it is overkill to find representative rules at all confidence levels, including, say, 0.1, 0.01, 0.001.

We analyze further properties of the confidence width to search for a faster computation. The key is to avoid much of the exploration in the naive algorithm by precomputing a small amount of side information in a single scan of the closures lattice. We explain now what side information would be sufficient; it is the same as used as a heuristic in [28] to compute a large subset of the representative rules faster[1]. The first step is to find out more about the rules $X' \to Y'$ that could be useful to compute the width of $X \to Y$.

Theorem 1. *Let $X \to Y$ (with $X \subset Y$, proper inclusion) be a representative rule for a fixed dataset $\mathcal{D}$ at some fixed values of support and confidence. Let $X' \to Y'$ be a different rule that makes it redundant, with $X' \subseteq Y'$, and assume $X' \to Y'$ has maximum confidence as in the definition of width. Then either $X = X'$ and Y' is a closed set, immediate superset of Y in the lattice of closed sets, and of maximum support among the closed supersets of Y; or else, $Y = Y'$, and X' is a minimal generator properly included in X and having minimum support among the proper subsets of X.*

Proof. First apply Lemma 1, but assume that we are in neither of the two cases, that is: $X' \subset X \subset Y \subset Y'$ where all the inclusions are proper. Consider the rules $X' \to Y$ and $X \to Y'$. Clearly, appealing again at Lemma 1, both make $X \to Y$ redundant as well. However, since Y is closed, $s(Y') < s(Y)$, and this implies that $c(X' \to Y') < c(X' \to Y)$; similarly, since X is a minimal generator, $s(X) < s(X')$, and again $c(X' \to Y') < c(X \to Y')$. Therefore, the confidence of $c(X' \to Y')$ is not maximum as required, and one of the two rules $X' \to Y$ and $X \to Y'$ will be the one having maximum confidence among those making $X \to Y$ redundant.   □

Now, the algorithmic alternative consists in modifying a closure lattice miner to maintain the side information we need. Since the resulting algorithm depends on which closed itemset miner is chosen as starting point, we cannot be fully explicit and keep generality at the same time: we just indicate the changes to be made into the closure miner. They are as follows: along the antimonotonicity-based construction of the frequent closures lattice and the minimal generators, we keep track of the largest existing support of the frequent closed supersets of each frequent closed set Y, let us denote it $mxs(Y)$. Similarly, for each minimal generator X, we keep track of the smallest existing support among the minimal generators properly contained in X, let us denote it $mns(X)$. Then the following proposition explains how to compute the width:

Proposition 2. *Consider a rule $X \to Y$, and assume that both $mxs(Y)$ and $mns(X)$ are defined. Then the width of $X \to Y$ is the minimum of the two values: $\frac{mns(X)}{s(X)}$ and $\frac{s(Y)}{mxs(Y)}$. If only one of $mxs(Y)$ and $mns(X)$ is defined, then the corresponding quotient gives the width.*

---

[1] The algorithm in [28], actually, may miss rules due to an incompleteness of the heuristic employed, caused by the fact that Property 9, as stated in that paper, is not true in all cases. This observation will be further elaborated in a later paper.

This follows directly from Theorem 1 since each of the two cases corresponds to one of the two options for a rule of maximum confidence making $X \rightarrow Y$ redundant. Note that we must compute $mns(X)$ for all minimal generators regardless of whether they are also closed, which is something that can happen (for instance, the empty set is often closed, and is always a minimal generator of the smallest closed set, possibly itself). Note also that some closures Y may not have frequent closed proper supersets, in the sense that all larger closures could fall below the support threshold; likewise, some minimal generators X, namely, the empty set, will lack minimal generators as proper subsets. For such cases, we leave $mxs(Y)$ and $mns(X)$ undefined. Rules where both are undefined do not have a confidence width value according to the definition, because no rule at all is able to make them redundant. Their width can be likened to "infinity". They have not arisen in our empirical analysis, probably due to the support threshold, and further theoretical development regarding this marginal case is undergoing.

Thus, algorithmically, we would use width by precomputing, at the time of finding closures from the dataset, or along the reading from a file if these are constructed by a separate closed set miner, the values $mxs(Y)$ for each frequent closed set Y and $mns(X)$ for each minimal generator X; then, for each representative rule $X \rightarrow Y$, we resort to Proposition 2 to compute $w(X \rightarrow Y)$ and use it either to filter (against a width threshold) or to sort the representative rules to be given as output.

Proposition 2 tells us also something else: we can discuss the confidence width according to two variants, one of them corresponding to a rule becoming redundant due to a larger consequent, and the other corresponding to a rule becoming redundant due to a smaller antecedent. It will be important shortly to take into account that the items discarded from the antecedent in this last case must still be present in the consequent, since we are assuming, as discussed immediately after Lemma 1, that right hand sides include left hand sides.

2.3   Squint-Based Threshold Setting

We propose here a way of connecting the confidence bound to the confidence width bound. The guiding intuition is as follows. First, we rephrase the confidence in a way that, informally, we call "squint": the extent to which we "see" small details. For squint q, sets that differ in a size ratio of q or higher will be considered distinguishable: their difference is actually seen. Note that this cannot be taken as a formal definition, since it may happen that one cannot distinguish X from Y nor Y from Z, yet X can be distinguished from Z. We take it just as an intuition.

Correlating the intuition of squint with the confidence threshold is easy. The implication $X \rightarrow Y$ (that is, $X \rightarrow XY$) is 100% true exactly if the set of transactions having X coincides with the (in principle potentially smaller) set of transactions having XY. Now, apply the guiding intuition for the squint parameter: if these sets can be distinguished, we discard the implication. For instance: at squint zero, sharpness is maximum, any existing difference is seen, and only absolute implications are accepted as association rules. However, at squint $q > 0$, to distinguish the set of transactions having X from those having XY we need that their difference has a size, relative to the larger of both sets, of at least q. That is: to distinguish enough counterexamples for the implication, $(s(X) - s(XY))/s(X)$ must be larger than q, and, conversely, the implication is accepted if $(s(X) - s(XY))/s(X) \leq q$, which is equivalent by straightforward algebraic manipulation to $s(XY)/s(X) = c(X \to Y) \geq (1 - q)$. (Note that this part also works for the case $q = 0$.) The confidence threshold corresponding to squint q is, then, $1 - q$.

Now, through a similar intuition, each of the two quantities, of which the smallest one provides the confidence witdh as per Proposition 2, can be connected to this "squint" parameter. To obtain a large enough width bound that allows us to "see" rule $X \to Y$, given squint q, the supports $s(X)$ and $mns(X)$ must be clearly different, and also the supports $s(XY)$ and $mxs(XY)$. Thus, we model the corresponding intuitions as $(mns(X) - s(X))/s(X) > q$ and $(s(XY) - mxs(XY))/mxs(XY) > q$. Straightforward algebraic manipulations lead, in both cases, to the condition $w(X \to XY) > 1+q$. That is, the confidence threshold, through the intuition associated to the squint parameter, provides us with a natural suggestion regarding how to set the threshold on the quantity under study: if the confidence threshold is $\gamma = 1 - q$, the natural first choice for confidence width threshold is $1 + q = 2 - \gamma$.

3   Blocked Rules

The main disadvantage often argued against confidence is as follows: for a threshold of, say, $2/3$ (or around 66%), consider a representative rule $A \to B$ of confidence slightly beyond the threshold. It is going to be provided as interesting in the output, suggesting that transactions having A tend to have also B. However, in case the actual frequency of B is rather high, say, 80%, the correlation is in fact negative, since B appears *less* often among the transactions having A than in the whole dataset. The natural reaction, consisting of a normalization by dividing the confidence by the support of B, gives in fact (an analogue to) the deviation from independence $s(AB)/s(A)s(B)$, also known as *interest*, *strength*, or *lift*, a natural measure that, however, lacks the ability to orient the rules, because, in it, the roles of A and B are absolutely symmetric, so that no preference is given for $A \to B$ versus $B \to A$. The same objection appears for the randomization-based proposal in [19]. Confidence width comes close to help but falls a bit short of offering a new solution to this problem. In this section, we relax slightly the notion of confidence width into a notion of "rule blocking" that progresses towards an alternative, nice solution to this difficulty.

For a specific motivating example, let us observe the outcome of mining for association rules at 5% support and 100% confidence the ADULT dataset, available at the UCI Repository [4]. The representative basis for these thresholds consists of 71 rules. In four of them, the consequent consists of the items "Male" and "Married-civ-spouse". In the other 67, the consequent is, in all cases, just "Male". For instance, we find "Craft-repair, Husband $\to$ Male" or "Husband, Some-college, United-States, White $\to$ Male".

Further examination reveals that all the left-hand sides consist of the item "Husband", together with one to four additional items. Domain knowledge suggests that all these 67 rules should be superseded by a single full-confidence rule "Husband → Male". However, tuple 7110 includes actually the item "Husband", and the item "Female" instead of "Male". Hence, such a rule does not appear due to the 100% confidence threshold and, instead, many rules that enlarge a bit the left-hand side (enough to avoid tuple 7110 so as to reach confidence 100%) show up. But the real information is the unsurprising (but reassuring) rule given by the domain knowledge and, to some extent, the fact that its confidence is below 100% due to the odd tuple; the user gains nothing by seeing 67 slightly different variations of the same fact. Confidence width does not help: the rule given by domain knowledge does not really make redundant, in the strict logical sense, the 67 rules mined, due to the extra items present in them. On the other hand, the presence of such a large family of rules, each of them improving the confidence only slightly over an existing rule, is a potentially very effective approach to outlier detection.

Yet another example, on the same dataset, that does not involve implications of full confidence but still allows for a similar argumentation, is the following rule, which relates family status with native country: "Unmarried → United-States", of rather high confidence (88%); it might be taken as a suggestion that people coming from abroad into the given U.S. community under analysis tend to come after marriage, but it may as well be an artifact due to the very large ratio of the sample that actually consists of U.S. natives, irrespective of their family status: over 89%. Note, however, that whereas this large support makes the high confidence of the rule "Unmarried → United-States" much less surprising, both high values carry related but different information: the distribution of the U.S. natives along the two different populations, the global one and the one of unmarried people, in principle could be different. The task is, then, to put the squint intuition into use in order to distinguish whether we should maintain both rules "Unmarried → United-States" and, so to say, "∅ → United-States" (the latter being essentially the same as to observe the high support of that item), because the slightly different information they carry is of interest, or we should consider the former subsumed by the latter. Note that either could have higher confidence than the other, depending on the dataset. To cater for such situations, we propose to work out a variation of confidence width, and a corresponding threshold obtained from the confidence threshold via the "squint" intuition, as follows.

3.1   Blocking a Rule with Another

Consider a rule $X \to Y$, and assume $X \cap Y = \emptyset$. We wish to discard it in case we find a rule $Z \to Y$, with $Z \subset X$, having almost the same confidence, and the task is to quantify this "almost". We propose to apply the squint intuition to compare the number of tuples having XY with the quantity that would be predicted from the confidence of the rule $Z \to Y$; if both sets of tuples are close enough in size, we keep $Z \to Y$ and forget about $X \to Y$. We will say, then, that Z "blocks" $X \to Y$.

Let $c(Z \to Y) = c$. If Y is distributed along the support of X at the same ratio as along the larger support of Z, we would expect $s(XY) \approx cs(X)$. We employ the "squint" intuition described in the previous section, and evaluate $X \to Y$ as follows:

Definition 3. *Given rule $X \to Y$, with $X \cap Y = \emptyset$, a proper subset $Z \subset X$ blocks $X \to Y$ at squint q if*

$$(s(XY) - c(Z \to Y)s(X))/(c(Z \to Y)s(X)) \leq q.$$

In case the difference in the numerator is negative, it would mean that $s(XY)$ is even lower than what $Z \to Y$ would suggest. If it is positive but the quotient is bounded by q, the difference is "not seen" and $X \to Y$ still does not bring high enough confidence with respect to $Z \to Y$ to be considered: it remains blocked. But, if the quotient is larger, and this happens for all Z, then $X \to Y$ becomes interesting since its confidence is higher enough than suggested by the rules of the form $Z \to Y$.

It can be readily checked that the particular problems of the ADULT dataset alluded to above are actually solved in this way. Namely, a bit of arithmetic with the actual supports in the dataset shows that, indeed, the rules given as example above, namely, "Craft-repair, Husband $\to$ Male", "Husband, Some-college, United-States, White $\to$ Male", or "Unmarried $\to$ United-States" get all blocked at minimally reasonable squint levels: only an extreme acuteness value for squint will be able to distinguish the different information provided by these rules from that of their blocking rules.

By way of comparison, note that we assume that redundancy due to larger consequents is handled by confidence width, whereas smaller antecedents only in some cases are handled appopriately by width, due to the stringent condition of logical consequence. With blocking, we handle similarly the case of smaller antecedents but in a way that is not as strict as logical consequence.

4 Empirical Validation

We describe first some experimentation made with the notion of confidence width. We compute closures using the C implementation provided by Borgelt [9]. On top of the obtained lattice of closures, we precompute the quantities mxs and mns as per the previous section at the time of loading the closures into our system, use hypergraph transversal techniques to find minimal generators [35], and thus obtain all the representative rules for the support and confidence bounds computed from the squint value. Table 1 indicates some parameters of the datasets on which we have tested our approach.

First, we consider two of the standard FIMI benchmarks [14], of very different characteristics: chess, which is a small but very dense dataset on which even high support constraints lead to huge amounts of closed sets and of rules, and the largish, much sparser dataset retail coming from a standard application domain (market basket analysis). We have computed the representative rules

Table 1. Dataset parameters

Dataset	Source	Transactions	Different Items
CHESS	FIMI	3196	75
RETAIL	FIMI	88162	16470
ADULT	UCI	32561	269
CMC	UCI	1473	36

and their widths, and we have plotted the number of rules passing each of a series of width thresholds. In all cases the computation has taken just a few seconds in a mid-range laptop.

If comparatively larger width values are expected to correlate in some sense with novelty, we wish the number of such rules above comparatively larger thresholds to decrease substantially. This is indeed the behavior we have found. With respect to the chess dataset, we have constructed rules of confidence 85% out of the closures lattice formed by frequent closed sets at support 80%. Even for such a large support, the number of closures is around 5083 and the representative rules amount to a number of 15067. It is known from the theoretical advances that all of them are fully irredundant, that is, omitting any of them loses information; however, it makes no sense to expect a human analyst to look at fifteen thousand rules.

We propose, instead, to look at the width values: for this dataset, they range in the quite limited interval between 1 and 1.22; and we see that if we impose a very mild bound of width above 1.005, only 2467 out of the 15067 rules reach it. This means that all the others, even if they are indeed irredundant, this is so due to a rather negligible confidence increase. Higher width bounds exhibit an interesting phenomenon of discontinuity, represented by each plateau of the graph in Figure 1 (left): the maximum confidence width of 1.22 is attained by two rules; a third comes close, and all three have high confidences (between 97% and 99%). Then seventeen more rules show up together near width 1.18, and nothing happens until the width bound gets below 1.13 where a bunch of 31 rules show up together. Below 1.11 we are again at a stable figure of 134 rules, and seventy more appear together at the already quite low confidence width bound of 1.075. All the others, up to 15067, have extremely low width. But the same role cannot be filled directly by confidence: the plot in Figure 1 (right) indicates that there are no steep decreases, no plateau suggesting a good cutpoint shows up, no hint that really any novelty is at play, and, above all, the following fact: the 51 rules of width 1.13 or more all have confidence of 90% or higher, but there are around 1950 other rules, of lower width, attaining the same confidence. Just width is able to focus on the 51 more novel ones.

With respect to retail, the behavior of the notion of width is very different, and also very interesting. Huge widths are reached: there are 18 rules whose width is beyond 560 (up to 855.94), whereas the highest next width is just 29: no rule has width between 29 and 560. Another plateau, at width 21, has 7 additional rules, and from there on the number of rules at each width threshold grows steadily.

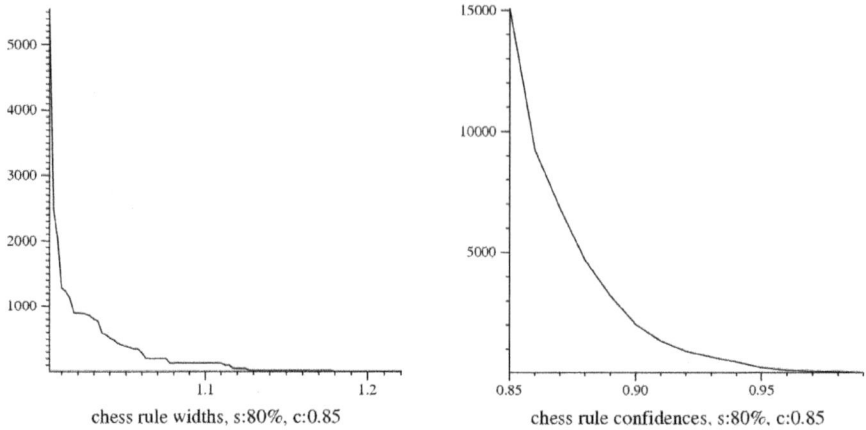

chess rule widths, s:80%, c:0.85   chess rule confidences, s:80%, c:0.85

Fig. 1. Chess: Number of rules per width and confidence

The FIMI datasets have all their items coded as opaque integers; therefore, the actual rules found cannot be intuitively assessed, in that we do not know their meaning. In order to understand better both the confidence width and the blocked rules, we have performed some further analysis of the very well-known ADULT dataset from the UCI Repository [4]. We use only the train data (we note that the test data has an extra dot in the class attribute). The numeric fields "fnlwgt", "capital-gain" and "capital-loss" were removed, as well as the field "education-num" which is fully redundant with the field "education". In fields "age" and "hours-per-week" the field name was concatenated to the numeric values, in order to distinguish which source to attribute to numeric items. No further cleaning or recoding was done. Table 2 shows the number of rules after the various filtering options. Each row in the table corresponds to a different value of the squint: all thresholds, including support and confidence, are computed from it and used consistently to get each of the figures. Confidence width is computed according to our proposal. Blocking, which in principle should be more powerful, is implemented here in a preliminary form: for each rule of large enough support and confidence, we just test whether it is blocked, at the given squint, by another rule that has also large enough support and confidence. See the Conclusions section for alternatives we wish to explore. The connection of support and squint is also very preliminary and under research: in this case we have used the support value $4 * q * r/M$ for squint q, where r is the average transaction size and M is the total number of items; but explanations and variants will be reported in future work.

The columns in Table 2 indicate the number of rules and the effect of filtering the representative rules through thresholds computed according to our proposal on the basis of the squint value. Their meaning is as follows:

Table 2. Number of rules in the ADULT dataset

Squint	Standard	Repr R	Block	Conf Wd	Both
0.10	7916	5706	1509	409	125
0.14	6518	4747	563	290	68
0.18	5270	3730	282	236	40
0.22	4289	2948	162	195	14
0.26	3641	2400	112	156	14
0.30	3024	2012	99	174	11
0.34	2740	1790	71	185	10
0.38	2547	1668	51	219	7
0.42	2255	1486	32	199	6
0.46	2056	1334	24	192	8
0.50	1865	1217	16	196	7

- Column "Standard" are the rules found by the standard apriori miner implementation [9]. We must mention that their number is less than the total number of rules since the Apriori rule miner employed only outputs rules with a single item in the consequent, as per the original proposals [3]; our system, and the rest of the figures, do not have this restriction.
- Column "Repr R" is the number of representative rules, which is optimum if we do not want to lose information.
- Column "Block" indicates the number of representative rules clearing the blocked rule condition.
- Column "Conf Wd" indicates the number of representative rules clearing the confidence width threshold.
- Column "Both" indicates the number of representative rules passing both constraints.

For the sake of arguing the interest of our process, we provide in Table 3 the *full* set of rules passing the thresholds at squint 0.32, with their supports, confidences, and confidence widths.

Whereas none is particularly surprising, the advantage is that now we know that, at the corresponding support, everything else is related to these rules through either redundancy, blocking, or lack of novelty; that each of these relationships can be quantified, and that in order to change the level up to which these relationships are computed it suffices to change a single parameter.

Finally, we have run our experiments also on an additional dataset: Contraceptive Method Choice, for which the results are displayed in Table 4. This dataset, abbreviated here CMC, is also from [4]; it is similar to ADULT in that it was originally conceived for a prediction task and in that it contains socioeconomic and demographic data where correlations among human factors can be potentially detected; but is very different in terms of size and density. Data come from an actual survey in Indonesia regarding demographic, religious, educational, and offspring data among women, run in 1987. Whereas in ADULT even the representative rules are long to explore manually, in this case the option clearly exists,

Table 3. Rules from the ADULT dataset filtered at squint 0.32

lhs	rhs	Support	Confid	C. Wd
∅	⇒ United-States,White	78.69%	78.69%	1.35
Husband	⇒ Male,Married-civ-spouse, United-States,White	33.91%	83.70%	1.56
Married-civ-spouse	⇒ Husband,Male, United-States,White	33.91%	73.74%	1.56
Not-in-family	⇒ ≤ 50K,United-States,White	18.06%	70.81%	1.36
Divorced	⇒ ≤ 50K,United-States,White	9.87%	72.32%	1.39
Black	⇒ ≤ 50K,United-States	7.62%	79.42%	1.42
hours-per-week:50	⇒ Male,United-States,White	6.37%	73.54%	1.4
Female,Some-college	⇒ ≤ 50K,United-States,White	6.06%	70.31%	1.37
Adm-clerical,Private	⇒ ≤ 50K,United-States,White	6.04%	69.43%	1.33
Self-emp-not-inc	⇒ Male,United-States,White	5.71%	73.20%	1.35
≤ 50K,Sales	⇒ Private,United-States,White	5.65%	68.95%	1.37

Table 4. Number of rules in the CMC dataset

Squint	Standard	Repr. Rules	Block filter	Conf Wd filter	Both filters
0.10	228	206	120	16	10
0.20	81	67	27	13	8
0.30	33	25	9	7	3
0.40	12	10	2	4	1
0.50	7	5	1	5	1

but it is a frustrating experience: two items ("Good-exposure-to-media", 92%, and "Wife-religion-islam", 85%) are prevalent to such an extent that almost all the rules have just one of these, or both, as consequent, and are therefore uninformative; "High-husband-education" follows closely (61%). Our approach points this out: the rule with empty antecedent "→ Good-exposure-to-media Wife-religion-islam" is clearly singled out beyond squint 0.35, and appears together with the rules "High-wife-education → Good-exposure-to-media High-husband-education" and "High-standard-of-living → Good-exposure-to-media High-husband-education" already at squint 0.25.

This exploration was fast (the closure space consisting of just 1863 closures, thus all rule computations taking just seconds on a mid-range laptop) and immediately suggests to proceed to a more acute exploration, of low squint, to see whether more benign thresholds for support, blocking, and confidence width (automatically compensated by a stricter confidence threshold) provide further information. From such a second phase, also fast, we just note that, at squint around 0.04, the representative rules are several hundred, but our automatically computed thresholds leave just around four dozen rules, most of which are now-unblocked variants with the three very frequent items as consequent (that can be readily discarded at a glance) plus the additional rule "No-children-so-far → No-contraceptive-method", missed in the previous exploration due to low width and support but having large confidence (almost 98%).

5 Conclusions and Further Work

We have proposed two objective approaches to the analysis of the novelty of association rules. A main intuition can be gleaned from the current early developments: it is known that, on the one hand, the standard support-and-confidence bound framework does a good preliminary job for avoiding statistical noise, but, on the other hand, fails somewhat to focus on the really interesting facts; and this is the main reason that has led to a flourishing of variants of notions of "implication degree" to replace confidence, blaming into it the problem. However, we consider now that a viable alternative is to leave the standard support-and-confidence setting on, and complement it, in order to gain further focus, with a measure that does not check the degree of the implications in an alternative way (thus, performing something intuitively analogous to confidence) but which checks a relative intensity of implication compared to the other rules mined in the same process.

Our proposals for this role are confidence width and a related form of blocked rules. Both compare rules among them in search of logical or intuitive redundancy: logical redundancy for the case of width, and a more relaxed, intuitive redundancy for blocking. Our experimental analysis is, admittedly, somewhat limited; but our work so far already suggests several interesting points. It shows that width has the ability to yield wide segments where a width threshold is very robust, and fixing it at a close but different value may select exactly the same rules. It tends to select rules of high confidence but is much more selective.

Also, our proposals open a door to a more human-centered development where one can find ways of evaluating this formal notion of novelty with respect to user-conceived naive notions of novelty. One potential development could be to design an interactive knowledge elicitation tool that, on the basis of the theory described here, could tune in, up to focusing on the user's intuitions for novelty, by showing a handful of unblocked rules of high width, asking the user to label them as novel or not novel: we should develop further the theory to take into account facts such as rules of high width (or support or confidence) being labeled as not novel, so that the labeling would have consequences on the values of these parameters for the rest of the rules.

We have proposed as well a rule mining framework in which, instead of asking the user to choose, with hardly any guidance, thresholds for all the parameters such as confidence, width, blocking thresholds, and possibly others, a single parameter is chosen, with a degree of semantic intuitive guidance, and then some of the necessary thresholds are autonomously computed by the system from that chosen value. In further work we will analyze the amenability of the support threshold to be treated in the same way; some of our experiments were done according to our preliminary results on that question.

Several major issues need further attention, and are described briefly in the next paragraphs.

5.1   Blocking Rules

Our current implementation only checks for blocking among the rules that have passed the support and confidence thresholds, that is, does not block all "block-able" rules; in most cases this is unproblematic thanks to the confidence width threshold. The analog of Proposition 1 is very easy to prove and we have incorporated this consideration into our experiments. Our further preliminary results suggest the use of a "double-confidence mining" approach analogous to the "double-support mining" approach described in [5], where it is shown how it can be advantageous, in order to distinguish representative rules under a support threshold, to mine closures above a milder threshold than the one set by the user, and employ that information to analyze redundancy of the rules above the user-set threshold. Further mathematical analysis of the formal properties of the blocking process is also necessary to clarify what is the sensible thing to do in case of "transitive blocking", whereby the blocking rule is itself blocked by a third rule: a case that may happen in practice and where we should study the properties we wish for the output rules.

5.2   Robustness

Other parameters instead of those described here may be manageable on the basis of the squint intuition, and possibly with potential advantages. Clearly a large family of candidates is given by the myriad of existing measures of intensity of implication (see [17], [18], [21], [37], among many others). But another family of parameters that could be employed are those whereby the family of closed sets is made more resilient, in the sense of tolerating a small degree of error and considering sets that are "almost" closed (see [8], [10], [11], and the references there). In that approach, the sort of analysis we perform on rules is made in the earlier stage of closure computation. It may become important to understand the potential advantages of this alternative.

In order to safely implement an exploration process as just described, ideally, the main parameter in a system like ours should offer, in as much as possible, *robustness* in some form of continuity: in most cases, slight modifications in the value of this parameter should not cause extremely big changes of the output. However, the very nature of the discrete sets we work with will impose occassionally abrupt changes. Whereas the squint intuition is already quite good in this respect, one potential way of improvement could be to use the "double-support mining" approach indicated above. The degree of robustness that this approach could contribute to the squint-based analysis is currently under study.

5.3   Alternative Ratios

The relative largeness of a set compared to another, as is considered along the squint-guided intuition, has one additional ambiguity. Namely: as denominator, in the ratio that relates each of our parameters to the squint intuition, we could have chosen the size of the other set. Then, the confidence threshold would result in $1/(1 + q)$ instead of $1 - q$, and the thresholds for the other parameters would end up using a factor $1/(1 - q)$ substituting for the uses of $(1 + q)$. We have chosen the milder, less restrictive, form of the bound for all the cases; further experimentation may suggest, for each threshold, either to stay with the milder bound, to replace it by the more strict one, or to somehow find a way of choosing among both options.

5.4   Revising the Closure Operator

The notion of representative rules is, in fact, only dependent on the dataset; however, the most efficient way to compute it is using the closure operator associated to the dataset. An alternative approach was suggested in [34] and [40], where similar approaches were proposed to treat separately the rules of confidence 100% from the rules of confidence at least γ (a minor variant of the same scheme, which reaches mathematically demonstrable absolute optimality of size for that approach, is described in [5]); all these variants are very tightly coupled to the closure operator, and are better than the representative rules when the confidence threshold is high and there are many rules of confidence 100%. In our preliminary tests we have not detected a major difference in the outcome from using representative rules or from using closure-based redundancy, but further analysis would be in order.

However, the closure operator itself is, essentially, the same mathematical object as the rules of confidence 100%; and, due to the blocked rules and the confidence width bound, we may as well be reluctant to employ anything related to them, since many of these rules may be some sort of artifacts, as we have already discussed in the case of the ADULT dataset. Therefore, we are left with a quandary: should we trust the closure operator when we distrust some of the full-confidence implications that conform it? The effect of this doubt on the representative rules is minor, since they are defined with no reference to closures nor implications and the role of the closure operator in their computation is, essentially, just algorithmic. For this reason, we have developed our approach in terms of representative rules, but further work is necessary to clarify to what extent bases constructed only from the closure space would offer better results.

5.5   Support Bounds versus Itemset Size

We have started to consider some natural heuristics for determining a support threshold. These are based on individual items; however, we can consider briefly here the option of setting different support thresholds for different itemset sizes. This simple idea has, in principle, a serious drawback: if one, generally, already lacks guidance to sensibly set a single support threshold, the problem is exacerbated if we are to set several of them, for the different itemset sizes. Our approach offers a way out: it is conceivable that the squint intuition can be used to suggest supports for different itemset cardinalities.

5.6 Item Distribution

In the computation of support, we have not distinguished among the various items. However, in practical cases, individual items may not be distributed uniformly; Zipfian-like laws or other distributions would be often natural. The effect of this consideration on the computation of the support bound has been discussed in [39]. The extensions of our approach to handle such cases are definitely worth further exploration.

Additional topics become open through our novel proposal: the applicability of the approach to outlier detection has been already hinted at; nowadays, pattern mining on structures more complex than itemsets is necessary in a wide spectrum of application areas, and exporting our approach may not be immediate; we definitely envision the possibility of applying this approach to preference analysis; and other application areas will call for additional developments.

Acknowledgements

For their supportive attitudes and helpful comments on the various versions of the various drafts on which this paper is based, the author is grateful to his research group LARCA at UPC, to the colleagues from neighbor groups that participate in the LARCA tasks, to the organizers of PAKDD'09 and the QIMIE workshop within it, to Dr. Gemma C. Garriga, and to the referees of three related conference or workshop submissions.

References

1. Aggarwal, C.C., Yu, P.S.: A New Approach to Online Generation of Association Rules. IEEE Transactions on Knowledge and Data Engineering 13, 527–540 (2001) (See also ICDE 1998)
2. Agrawal, R., Imielinski, T., Swami, A.: Mining Association Rules Between Sets of Items in Very Large Databases. In: SIGMOD 1993, pp. 207–216 (1993)
3. Agrawal, R., Mannila, H., Srikant, R., Toivonen, H., Verkamo, A.I.: Fast Discovery of Association Rules. In: Fayyad, U., et al. (eds.) Advances in Knowledge Discovery and Data Mining, pp. 307–328. AAAI Press, Menlo Park
4. Asuncion, A., Newman, D.J.: UCI Machine Learning Repository. University of California, School of Information and Computer Science, Irvine, CA (2007), http://www.ics.uci.edu/~mlearn/MLRepository.html
5. Balcázar, J.L.: Redundancy, Deduction Schemes, and Minimum-Size Bases for Association Rules (submitted for publication), Pascal Report 4259: http://eprints.pascal-network.org/archive/00004259; http://www.lsi.upc.edu/~balqui/papers.html
6. Baralis, E., Chiusano, S., Garza, P.: On Support Thresholds in Associative Classification. In: ACM Symp. on Applied Computing, pp. 553–558 (2004)
7. Bayardo, R., Agrawal, R., Gunopulos, D.: Constraint-Based Rule Mining in Large, Dense Databases. In: ICDE 1999, pp. 188–197 (1999)
8. Besson, J., Robardet, C., Boulicaut, J.-F.: Mining Formal Concepts with a Bounded Number of Exceptions from Transactional Data. In: Goethals, B., Siebes, A. (eds.) KDID 2004. LNCS, vol. 3377, pp. 33–45. Springer, Heidelberg (2005)

9. Borgelt, C.: Efficient Implementations of Apriori and Eclat. In: Workshop on Frequent Itemset Mining Implementations (2003), `borgelt.net`
10. Boulicaut, J.-F., Bykowski, A., Rigotti, C.: Free-Sets: A Condensed Representation of Boolean Data for the Approximation of Frequency Queries. Data Min. Knowl. Discov. 7(1), 5–22 (2003)
11. Calders, T., Rigotti, C., Boulicaut, J.-F.: A Survey on Condensed Representations for Frequent Sets. In: Boulicaut, J.-F., De Raedt, L., Mannila, H. (eds.) Constraint-Based Mining and Inductive Databases. LNCS (LNAI), vol. 3848, pp. 64–80. Springer, Heidelberg (2004)
12. Ceglar, A., Roddick, J.F.: Association Mining. ACM Computing Surveys 38 (2006)
13. Coenen, F., Leng, P., Zhang, L.: Threshold Tuning for Improved Classification Association Rule Mining. In: Ho, T.-B., Cheung, D., Liu, H. (eds.) PAKDD 2005. LNCS (LNAI), vol. 3518, pp. 216–225. Springer, Heidelberg (2005)
14. Frequent Itemset Mining Implementations Repository, `http://fimi.cs.helsinki.fi/`
15. Guigues, J.-L., Duquenne, V.: Famille Minimale d'Implications Informatives Résultant d'un Tableau de Données Binaires. Math. et Sci. Humaines 24, 5–18 (1986)
16. Ganter, B., Wille, R.: Formal Concept Analysis. Springer, Heidelberg (1999)
17. Garriga, G.C.: Statistical Strategies for Pruning All the Uninteresting Association Rules. In: ECAI 2004, pp. 430–434 (2004)
18. Geng, L., Hamilton, H.J.: Interestingness Measures for Data Mining: a Survey. ACM Comp. Surveys 38 (2006)
19. Guillaume, S., Guillet, F., Philippé, J.: Improving the discovery of association rules with intensity of implication. In: Żytkow, J.M. (ed.) PKDD 1998. LNCS, vol. 1510, pp. 318–327. Springer, Heidelberg (1998)
20. Hájek, P., Holeňa, M.: Formal Logics of Discovery and Hypothesis Formation by Machine. Theoretical Computer Science 292, 345–357 (2003)
21. Hébert, C., Crémilleux, B.: A Unified View of Objective Interestingness Measures. In: Perner, P. (ed.) MLDM 2007. LNCS (LNAI), vol. 4571, pp. 533–547. Springer, Heidelberg (2007)
22. Kanimozhi Selvi, C.S., Tamilarasi, A.: Association Rule Mining with Dynamic Adaptive Support Thresholds for Associative Classification. In: IEEE Int. Conf. Comput. Intell. and Multimedia Appl., pp. 76–80 (2007)
23. Kautz, H., Kearns, M., Selman, B.: Horn approximations of empirical data. Artificial Intelligence 74, 129–145 (1995)
24. Kawahara, M., Kawano, H.: Mining Association Algorithm with Threshold Based on ROC Analysis. In: IEEE Hawaii International Conference on System Sciences (2001)
25. Khardon, R., Roth, D.: Reasoning with models. Artificial Intelligence 87, 187–213 (1996)
26. Kryszkiewicz, M.: Representative Association Rules. In: Wu, X., Kotagiri, R., Korb, K.B. (eds.) PAKDD 1998. LNCS, vol. 1394, pp. 198–209. Springer, Heidelberg (1998)
27. Kryszkiewicz, M.: Fast discovery of representative association rules. In: Polkowski, L., Skowron, A. (eds.) RSCTC 1998. LNCS (LNAI), vol. 1424, pp. 214–221. Springer, Heidelberg (1998)
28. Kryszkiewicz, M.: Closed Set Based Discovery of Representative Association Rules. In: Hoffmann, F., Adams, N., Fisher, D., Guimarães, G., Hand, D.J. (eds.) IDA 2001. LNCS, vol. 2189, pp. 350–359. Springer, Heidelberg (2001)

29. Liu, B., Hsu, W., Ma, Y.: Pruning and Summarizing the Discovered Associations. In: KDD 1999, pp. 125–134 (1999)
30. Luxenburger, M.: Implications Partielles dans un Contexte. Math. et Sci. Humaines 29, 35–55 (1991)
31. Ma, S., Hellerstein, J.L.: Mining Mutually Dependent Patterns for System Management. IEEE Journal on Selected Areas in Communications 20, 726–734 (2002)
32. Megiddo, N., Srikant, R.: Discovering Predictive Association Rules. In: KDD 1998, pp. 274–278 (1998)
33. Omiecinski, E.R.: Alternative Interest Measures for Mining Associations in Databases. IEEE Trans. on Knowledge and Data Engineering 15, 57–69 (2003)
34. Pasquier, N., Taouil, R., Bastide, Y., Stumme, G., Lakhal, L.: Generating a Condensed Representation for Association Rules. Journal of Intelligent Information Systems 24, 29–60 (2005)
35. Pfaltz, J.L., Taylor, C.M.: Scientific Discovery through Iterative Transformations of Concept Lattices. In: Workshop Discr. Math. and Data Mining, SDM 2002, pp. 65–74 (2002)
36. Phan-Luong, V.: The Representative Basis for Association Rules. In: ICDM 2001, pp. 639–640 (2001)
37. Tan, P.-N., Kumar, V., Srivastava, J.: Selecting the Right Objective Measure for Association Analysis. Inf. Syst. 29(4), 293–313 (2004)
38. Wild, M.: A Theory of Finite Closure Spaces Based on Implications. Adv. Math. 108, 118–139 (1994)
39. Xiong, H., Tan, P.-N., Kumar, V.: Mining Strong Affinity Association Patterns in Data Sets with Skewed Support Distribution. In: Third IEEE International Conference on Data Mining, ICDM 2003, p. 387 (2003)
40. Zaki, M.: Mining Non-Redundant Association Rules. Data Mining and Knowledge Discovery 9, 223–248 (2004)
41. Zaki, M., Ogihara, M.: Theoretical foundations of association rules. In: Workshop on research issues in DMKD (1998)

PAKDD Data Mining Competition 2009: New Ways of Using Known Methods

Chaim Linhart[1], Guy Harari[1], Sharon Abramovich[2], and Altina Buchris[2]

[1] School of Computer Science, Tel Aviv University, Tel Aviv 69978, Israel
chaiml@post.tau.ac.il
[2] Department of Statistics and Operations Research, Tel Aviv University,
Tel Aviv 69978, Israel

Abstract. The PAKDD 2009 competition focuses on the problem of credit risk assessment. As required, we had to confront the problem of the robustness of the credit-scoring model against performance degradation caused by gradual market changes along a few years of business operation. We utilized the following standard models: logistic regression, KNN, SVM, GBM and decision tree. The novelty of our approach is two-fold: the integration of existing models, namely feeding the results of KNN as an input variable to the logistic regression, and re-coding categorical variables as numerical values that represent each category's statistical impact on the target label. The best solution we obtained reached 3rd place in the competition, with an AUC score of 0.655.

Keywords: data mining, logistic regression, KNN, credit risk assessment.

1 Introduction

The offer of credit for potential clients is a very important service for stimulating consumption in the market. One main difficulty credit scoring modelers have to contend with is gradual market changes which occur during the collection of data. This difficulty increases the risk when the credit is lent for long term payment.

The PAKDD 2009 data mining competition focused on the model's robustness against performance degradation caused by market gradual changes along several business years [1]. We participated in this competition as part of the requirements of the Data Mining course given by Professor Yoav Benjamini in Tel-Aviv University.

The challenge was as follows. We were given three datasets, which were collected over different years and consist of 30 explanatory variables and one binary target variable. The first dataset, which was used for model selection, is labeled and contains 50,000 samples collected during 2003. The second dataset consists of 10,000 unlabeled samples collected during 2005 and was used for model evaluation. After selecting a model we could apply it on this dataset, submit the results to the leaderboard web-site, and compare its performance to the scores attained by other teams. The third dataset, called the prediction data, consists of 10,000 unlabeled samples from 2008, and was used for grading the performance of the final models of all competitors. Performance of each model was evaluated by area under ROC curve (AUC, in short).

T. Theeramunkong et al. (Eds.): PAKDD Workshops 2009, LNAI 5669, pp. 99–105, 2010.
© Springer-Verlag Berlin Heidelberg 2010

2 Data Preparation

Initial observations revealed that some of the explanatory variables are not useful for analysis, since they are constant in either the modeling or prediction data. A small number of samples in the modeling dataset have unreasonable or missing values, so we ignored them. We replaced unreasonable and missing values in the prediction data, as detailed below. We also tried to remove samples with area and profession codes that are absent from the leaderboard or prediction data. This gave better results on the modeling data, but performed worse on the leaderboard dataset, so we abandoned this approach.

We noticed that some variables have a significantly different distribution in the modeling data than in the leaderboard and/or prediction data. For example, *AREA_CODE_RESIDENCIAL_PHONE* is "50" in 22%, 5%, and 15% of the samples in the modeling, leaderboard, and prediction datasets, respectively. Another example is *PAYMENT_DAY*, which receives the value "15" in only 0.2% of the modeling samples, and 21% of the prediction samples. These differences might lead to degraded performance on the prediction data – the models are fitted to data with certain characteristics, and tested on data with different distributions.

Numerical variables: Unreasonable values, such as age 0 or extremely high income, were replaced by the median value of the corresponding variable. In order to account for possible changes in the value of the local currency over time (e.g., due to inflation), we standardized the two income variables to mean 0 and standard deviation 1. We also experimented with other transformations, such as logarithm and square root.

We noticed that some samples contain 0 in the income variables *PERSONAL_NET_INCOME* and *MATE_INCOME*. The distribution of the target variable suggests that at least some of these values do not really represent zero income. For example, when *PERSONAL_NET_INCOME* is 0, the target variable is 0 in 83% of the cases; when the income is 50-150, it's 77%; for 150-250, it's 76%; and for 350-450 (approximately the mean income) it's 79%. This suggests that a value of zero indicates either no income, or a missing value. Therefore, as with several other variables, we replaced the 0's by the mean value (not including 0's).

Textual variables: We replaced the two personal reference textual variables by a single numerical variable that holds the sum of their lengths. This was done since we discovered a relationship between the length of the personal references and the target variable.

Categorial variables: Boolean variables and categorical variables with a small number of categories (such as *MARITAL_STATUS*) are easily handled by all the models we applied – each category is replaced by several boolean indicator variables, one per category. Variables with a large number of categories, such as *ID_SHOP* and *PROFESSION_CODE*, pose a difficult challenge. We first added indicator variables for the most frequent values of each such variable. However, using many such variables in a logistic regression, for example, is prone to over-fitting. On the other hand, using only a small number of the indicator variables utilizes the information in the corresponding categories while effectively ignoring the information in the rest of the categories. We thus developed a method for transforming these categorial variables into numerical variables in a similar approach taken in [2]. These variables are called here "*P-VAL* variables" and are described in what follows.

The target variable *TARGET_LABEL_BAD* gets the values 0 (good) and 1 (bad) in 40,105 and 9,868 (legal) modeling samples, respectively. Given a categorial variable *X*, we compared the distribution of *TARGET_LABEL_BAD* in each category of *X* to that of the entire data. We tried the following three transformations:

I.   "Probs": The proportion of 0's (good clients) among each category, that is, for a sample with *X=c* we replaced the category *c* by the fraction of 0's in *TARGET_LABEL_BAD* among all the samples with *X=c*.

II. "P-values": The probability to obtain at least/most the observed number of *TARGET_LABEL_BAD*=0 in a category, given the total number of 0's and 1's in *TARGET_LABEL_BAD*. Assume there are *K* samples with *X=c*, out of which K_1 have *TARGET_LABEL_BAD*=1 and K_0 have *TARGET_LABEL_BAD*=0. We can view these *K* samples as a series of samples from the whole set of samples without replacement, and thus we may use the hypergeometric distribution to test whether the *K* samples were drawn randomly from the entire set. We use a two-sided test to detect a tendency both to 0 and to 1. In order to preserve this information in the numerical variable, we replaced categories with a tendency to 0 by the above *p*-value, and categories with many 1's by one minus the *p*-value.

III. "Logit": As in II, but taking the logit of the *p*-value for categories with over-representation of 0's in *TARGET_LABEL_BAD*, and taking –logit(*p*-value) for categories with tendency to 1's. Thus, categories with a similar 0/1 distribution to that of the entire dataset, as well as very rare categories (that are present in only a couple of samples), are replaced by values close to 0. Categories in which there are statistically many samples with *TARGET_LABEL_BAD*=0 are replaced by very small (negative) values. Likewise, categories with a strong tendency for *TARGET_LABEL_BAD*=1 are replaced by large (positive) values.

Table 1. Main pre-processing steps performed on the data

The problem/issue	Variables involved	Our solution
Constant in modeling or prediction data	QUANT_BANKING_ACCOUNTS FLAG_MOBILE_PHONE FLAG_CONTACT_PHONE COD_APPLICATION_BOOTH FLAG_CARD_INSURANCE_OPTION FLAG_OTHER_CARD QUANT_DEPENDANTS EDUCATION	Omit variables
Illegal values in modeling data	SEX	Remove samples with illegal values
Unreasonable values	AGE MONTHS_IN_RESIDENCE PERSONAL_NET_INCOME MATE_INCOME	Replace unreasonable values with the median of the variable

Table 1. (*continued*)

Different categories and distribution of values in model and prediction datasets	SHOP_RANK	Omit the variable
Categorial variable with many categories	AREA_CODE_RESIDENCIAL_PHONE PROFESSION_CODE ID_SHOP PAYMENT_DAY	1. Create indicator variables for the most frequent categories 2. Transform to numerical variables using one of three methods: Probs, P-values, Logit
Currency changes over time (inflation)	PERSONAL_NET_INCOME MATE_INCOME	Standardize the variables to mean 0 and std 1
Textual variables	PERSONAL_REFERENCE_1 PERSONAL_REFERENCE_2	Replace by sum of lengths

3 Modeling

We used the cross validation approach (5-fold CV) to estimate the performance of our models. Note, however, that since the modeling, leaderboard and prediction datasets were not sampled from the same distribution, better performance on the modeling data does not guarantee improved results on the other two datasets.

Logistic model: We fitted logistic models using the glm() function in R [3], starting with single variables, and went on to include interactions between variables. We found that using our transformation of categorial variables into numerical variables solves the difficulty of ranking the categories - which is necessary for a monotonous relation, as the one the model tries to fit.

KNN: We implemented our own KNN function. For each test sample, it first identifies the training samples with: (a) The same sex, (b) The same marital status, (c) A similar age (ages different by less than some predefined threshold), and (d) A similar income (salaries that are bounded from both sides by some pre-defined multiplicative factor). It then computes the distance between the test sample and each of these training samples, using different weights for the various variables. It is worth mentioning that different types of variables require a different distance metric. For numeric variables we used the Euclidean distance, whereas categorial variables got a zero weight when levels were equal and some positive weight otherwise. Finally, the procedure reports the fraction of the k nearest neighbors with *TARGET_LABEL_BAD=1*, as well as the logit of its p-value (as described above for the "P-VAL" variables).

Logistic + KNN combined model: We combined the KNN and logistic models by feeding the results of KNN as input to the logistic model. In other words, we added two new variables, called *KNN_PROBS* and *KNN_PROBS_PVALS*, that contained the results of our KNN procedure in "Probs" and "Logit" transformation, respectively (notice that the KNN procedure was executed on both the training and test sets, in order to obtain the value of the two aforementioned variables for all samples – the training samples, to which the logistic model is fitted, and the test samples, on which it is tested).

We also experimented with several other models, such as decision tree, SVM (support vector machines) and GBM (generalized boosted models) as implemented in R [3]. However, they did not yield good results.

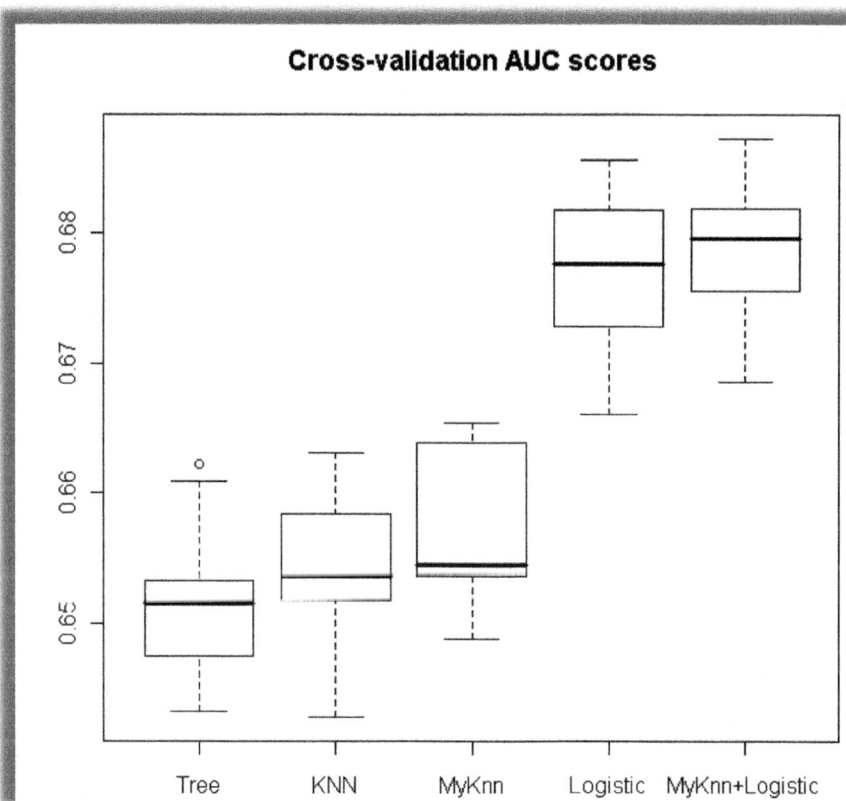

Fig. 1. AUC scores of the main models we studied. Scores were obtained using two iterations of 5-fold cross validation tests on the entire modeling dataset. "MyKnn" refers to our implementation of KNN. "MyKnn + Logistic" is the combined model.

4 Results

Logistic model: The logistic model gave reasonable results on the leaderboard data. Once we included the "P-VAL" variables, the score improved further. Interestingly, when we tested which type performs best, the results of the cross validation procedure indicated that the transformation of type "Probs" outperforms the others. However, the "Logit" transformation yielded the best score on the leaderboard dataset. A possible explanation is that the exact ratio of 0's and 1's in the target variable change over time, whereas statistically significant tendencies do not. The best logistic model we obtained reached an AUC score of 0.677 on the modeling data (see Figure 1), and 0.6125 on the leaderboard data.

An important observation is that a logistic model with many variables tended to return degraded results on the leaderboard data, even though it improved the results as assessed by the cross validation procedure on the modeling data. This might indicate an over-fitting of the model to specific characteristics of the modeling data, which change over time (recall that the leaderboard samples are two years after the modeling samples).

KNN: Our KNN procedure with $k=250$ attained an AUC score of 0.654 on the modeling data (Fig. 1) – less promising that the logistic model. However, the two models received the same score on the leaderboard data. Surprisingly, this was achieved by our KNN implementation using cutoffs and weights that were set by mere intuition on which variables are more important for predicting the target variable. Due to lack of time, we did not implement any procedure for optimizing the parameters of the KNN model. However, based on a couple of experiments, we believe that small changes to these parameters have very little effect on the results.

Logistic + KNN combined model: Combining the two models, as explained above, gave the best results. Our final logistic model consisted of 43 variables, including two variables that contained the results of our KNN procedure (*KNN_PROBS* and *KNN_PROBS_PVALS*), seven "P-VAL" variables (of type "Logit") and two indicator variables for frequent categories (area code 31 and profession 950); the rest of the variables were original variables (after the transformations we applied) and interactions between several pairs of variables (e.g., all pairwise interactions of *AGE*, *SEX*, and *MARITAL_STATUS*). The AUC score of the final model is 0.68 on the modeling data, 0.6177 on the leaderboard data and 0.655 on the prediction data – which is ranked 3rd in the competition.

5 Conclusions and Summary

We conclude that both KNN and logistic models describe the data quite well. However, these results may be misleading since the long execution time of KNN compelled us to attempt it with very few combinations of parameters and variables. Also, since we have limited experience with SVM and GBM, we cannot conclude whether they can or cannot model the data in the competition as well as KNN and logistic regression. Interestingly, the logistic model attained higher scores than the KNN approach in the CV test on the modeling data (see Figure 1), but both methods performed equally well on the leaderboard data, indicating perhaps that the logistic model is more over-fitted to the characteristics of the modeling data than KNN. Combining KNN with a logistic model gave the best results in our experiments.

Some variables that may have some influence on the target variable were omitted from our analysis for technical reasons, such as different names of categories between the modeling and prediction data. Replacement of missing or unreasonable values could be performed by a more suitable procedure, such as maximum-likelihood based methods.

We believe that the method we described for transforming categorial variables into numerical variables, as well as our combination of KNN with logistic regression, are interesting and could be applied on other datasets. Another interesting approach could be feeding the logistic model with results from other models, such as SVM or neural networks. Due to lack of time, we paid little attention to the issue of feature selection, which could have enhanced the performance of our models.

Acknowledgments. We gratefully acknowledge the help, support and advice of Prof. Yoav Benjamini from the department of Statistics and Operations Research at Tel Aviv University. The presentation of this work was supported by the Edmond and Beverly Sackler Chair in Bioinformatics, the school of Computer Science, and the Sackler Faculty of exact sciences, all in Tel Aviv University.

References

1. PAKDD data mining competition 2009, Credit risk assessment on a private label credit card application (2009), http://sede.neurotech.com.br/PAKDD2009
2. Ritchie, M.D., Hahn, L.W., Roodi, N., Bailey, L.R., Dupont, W.D., Parl, F.F., Moore, J.H.: Multifactor-dimensionality reduction reveals high-order interactions among estrogen-metabolism genes in sporadic breast cancer. Am. J. Hum. Genet. 69(1), 138–147 (2001)
3. R Development Core Team. R: A language and environment for statistical computing. R Foundation for Statistical Computing, Vienna, Austria (2009), http://www.R-project.org

Feature Selection for Brain-Computer Interfaces

Irena Koprinska

School of Information Technologies, University of Sydney, Sydney NSW 2006, Australia
irena@it.usyd.edu.au

Abstract. In this paper we empirically evaluate feature selection methods for classification of Brain-Computer Interface (BCI) data. We selected five state-of the-art methods, suitable for the noisy, correlated and highly dimensional BCI data, namely: information gain ranking, correlation-based feature selection, ReliefF, consistency-based feature selection and 1R ranking. We tested them with ten classification algorithms, representing different learning paradigms, on a benchmark BCI competition dataset. The results show that all feature selectors significantly reduced the number of features and also improved accuracy when used with suitable classification algorithms. The top three feature selectors in terms of classification accuracy were correlation-based feature selection, information gain and 1R ranking, with correlation based feature selection choosing the smallest number of features.

Keywords: brain-computer interfaces, classification of EEG data, information gain ranking, correlation-based feature selection, ReliefF, consistency-based feature selection, 1R ranking.

1 Introduction

A BCI is a system which allows a person to control devices such as a computer cursor or robotic limb by only using his/her thoughts. It aims to help severely paralyzed people to communicate by providing a way which doesn't depend on muscle control but only on their thoughts. Building BCIs is an interdisciplinary field combining expertise in medicine, neurology, psychology, machine learning, statistics and signal processing. It has been a very active area of research in the last 15 years, stimulated by new understanding of the brain function and EEG signals, the availability of powerful and low cost computer equipment and the wider recognition of the needs of people with severe neuromuscular disorders [1, 2].

BCI systems are based on recording EEG brain activity and recognizing patterns associated with mental tasks. It is known that mental tasks such as imagining a movement of the right and left hand are associated with patterns of EEG activity in the left and right side of the motor cortex, respectively. These patterns are associated with various changes in EEG activity. For example, the mu rhythm (8-12 Hz) and beta rhythms (18-26 Hz) are known to decrease during movement or preparation for the movement (event-related desynchronization) or increase after movement (event-related synchronization) [2]. It is possible to select a small set of mental tasks that activate different parts of the brain to make the recognition easier. Then, supervised

T. Theeramunkong et al. (Eds.): PAKDD Workshops 2009, LNAI 5669, pp. 106–117, 2010.

classification algorithms are employed to learn to recognize these patterns of EEG activity, i.e. to learn the mapping between the EEG data and the classes corresponding to mental tasks [3].

From data mining point of view this is a challenging task for several reasons. Firstly, the EEG data is noisy and correlated as many electrodes are fixed on the small scalp surface and each electrode measures the activity of thousands of neurons [4]. In addition, the quality of the data is affected by the different degree of attention of the subject and changes in their concentration during the data recording; these factors introduce additional noise. Secondly, the dimensionality of the data is high as many channels are recorded and several features are extracted from them [3]. At the same time the number of training examples is small as collecting labelled data is time consuming and cognitively demanding process for the subjects.

In this paper we focus on feature selection to address these challenges of BCI data, namely the noisy, correlated and highly dimensional data, with a small number of training examples. Feature selection is the process of removing irrelevant and redundant features and selecting a small set of informative features that are necessary and sufficient for good classification. It is one of the key factors affecting the success of a classification algorithm. Feature selection also reduces the dimensionality of data which means faster building of the classifier and often producing more compact and easier to interpret classification rule [5]. Furthermore, it is needed to avoid the above mentioned curse of dimensionality problem - small ratio of sample size to number of features.

The main goal of our study is to empirically evaluate a number of state-of-the-art feature selection methods for classification of BCI data. Comprehensive surveys of feature selection for classification can be found in [6] and [7]. An empirical comparison of feature selection methods on UCI benchmark datasets was presented in [8]. A brief survey of machine learning techniques, including feature selection methods, that can be applied to BCI data is given in [9]. In contrast, our goal is to empirically compare five important feature selection methods on benchmark BCI data from the BCI competition, which hasn't been done before. The five methods we chose - Information Gain Ranking (IG), Correlation-Based Feature Selection (CFS), ReliefF, Consistency-Based Feature Selection (Consistency) and 1R Ranking (1RR) - are state-of-the-art feature selectors, have been successfully applied in other domains and are appropriate for the nature of the EEG data. Only one of them, CFS, has been previously applied for classification of BCI data in our recent study [10].

In addition, we also evaluate a number of classification algorithms with these feature selection methods. A variety of algorithms have been applied in BCI systems, e.g. linear classifiers [3, 9, 11] which are still the favorite approach, neural networks [12], nearest neighbor classifiers [11] and support vector machines [4]. Lotte et al [3] survey classification algorithms for BCI data and note that it is hard to compare them as the experimental setup, preprocessing and feature selection are different in the reported studies. Hence, we also contribute to the evaluation of classifiers, using a benchmark dataset, the same pre-processing and the same feature selection methods.

The next section briefly describes the feature selection methods we compare. Section 3 presents the dataset, pre-processing and experimental methodology. The results are presented and discussed in Section 4. Section 5 concluded the paper.

2 Feature Selection Methods

We chose five state-of-the-art feature selection methods: IG, CFS, ReliefF, Consistency and 1RR. All of them are examples of filter methods for feature selection [13]. The distinction between filter and wrapper methods for feature selection is based on their connection with the classification algorithm. Filters evaluate and rank features or feature subsets prior to learning and independently of the classification algorithm. Wrappers evaluate and rank feature subsets for a particular target classification algorithm. They work well as the feature selection is tuned for the particular classifier but are also very slow as a classifier needs to be built for every subset and evaluated using cross validation. Due to the large number of features in our task, the application of wrappers was not feasible in this study.

Feature selection methods can also be categorized based on what they evaluate and rank: individual features or subsets of features. CFS and Consistency evaluate subset of features and produce a single feature subset; IG, Relief and 1RR evaluate all features individually and rank them; a feature subset selection is achieved by selecting the highest N ranked features or all features with a value above t, where N and t are user-specified thresholds.

IG, CFS, ReliefF and Consistency were included in the comparison of 6 feature selection methods on 15 benchmark and 3 large datasets in [8], which also included principle component analysis and wrapper. When the speed was not an issue, the wrapper was found to be the best performing method in terms of accuracy; otherwise CFS, Consistency and ReliefF were the best. The evaluation was conducted using only 2 common classification algorithms (decision trees and naïve Bayes) while in this study we use 10 algorithms, see Section 3.2.

IG. This is a very popular and successful feature selection method for high dimensional data, widely used in the area of text classification [14]. Given a set of classes $C = \{c_1,...,c_k\}$, the information gain of a feature f, $IG(f)$, is the expected reduction in entropy H caused by observing f:

$$IG(f) = H(C) - H(C \mid f), \text{ where}$$

$$H(C) = -\sum_{i=1}^{k} P(c_i) \log P(c_i), \ H(C \mid f) = -P(f)\sum_{i=1}^{k} P(c_i \mid f) \log P(c_i \mid f).$$

The computation is done for each feature across all classes and then the features are ranked based on their IG value; the higher the value the more informative the feature is. To select the top N features, we experimented with different thresholds and report the best results which were achieved for $t = 0$.

CFS. CFS is a simple and fast feature subset selection method developed by Hall [5]. It searches for the "best" subset of features where "best" is defined by a heuristic which takes into consideration two criteria: 1) how good the individual features are at predicting the class and 2) how much they correlate with the other features. Good subsets of features contain features that are highly correlated with the class and uncorrelated with each other. Thus, CFS directly handles correlated and irrelevant features, which makes it suitable for EEG data. The search space is very big for employing a brute-force search algorithm. We used the best first (greedy) search option starting with an empty set of features and adding new features.

ReliefF. Relief [15] is an instance-based feature ranking method for two-class problems. ReliefF [16] is an extension of Relief for multiclass problems. Relief ranks the features based on how well they distinguish between instances that are near to each other. It randomly selects an instance Ri from the data and finds the nearest neighbor H from the same class and the nearest neighbor M from the other class. Then it updates the quality score of each feature by comparing the feature values of Ri with H and M. If Ri and H have different values of f, this means that two instances from the same class are separated by f (not desirable), the score of f is decreased. If Ri and M have different values of f, this means that two instances from different classes are separated by f (desirable), the score of f is increased. The process is repeated for m randomly selected instances. ReliefF is also more robust that Relief as it uses k nearest neighbors. We used $k = 10$ and $m = $ all instances, i.e. all instances in the training data were sampled which increases the reliability of the feature scores.

ReliefF is very appropriate for EEG data as it works well on noisy and correlated features and scales well for high dimensional data due to its linear time complexity. Similarly to IG, ReliefF ranks all features and requires a threshold t for selecting the top N features. We report the best results which were achieved for $t = 0.05$.

Consistency. It selects a subset of features by searching the space of subsets guided by a class consistency measure [17]. More specifically, it looks for combinations of features that are mainly associated with the same class. Initially, the best subset consists of all features and the consistency threshold is set to 0. If the candidate subset has a better class consistency score and less or equal number of features than the current one, it becomes the best subset. We used best first search as a search method. Consistency is a fast algorithm, able to identify dependency between features [8].

1RR. 1RR [18] is based on the 1R classification algorithm [19]. 1R generates a classification rule (1-rule) that tests the values of a single feature, i.e. it generates a one-level decision tree. It does this by creating a 1-rule for all features and then selecting the one with the highest classification accuracy. Holte [19] shows that the simple 1R classifier compares favourably with state-of-the art classifiers on standard machine learning datasets and explains this with the rudimentary structure of many real-world datasets, which motivates the use of simple algorithms first. 1RR is an extension of the 1R algorithm and is used for feature selection. It ranks all features based on the classification accuracy of their 1-rules and then selects the top N features based on a value threshold t. Thus, it is based on the assumption that the accuracy of each feature is an indicator of its relevance. While 1R can be seen as a method for selecting 1 feature, 1RR is used to select sub-set of features. 1RR is a simple and fast algorithm and was shown to be an effective feature selector for document classification [20]. In our experiments we used a cut-off threshold $t = 40$.

3   Experimental Methodology

3.1   Data and Preprocessing

We used dataset IIIa from the latest BCI competition, BCI III [21]. It contains re-cordings for three subjects (K3b, K6b and L1b) in a four-class classification problem.

We briefly summarized the data acquisition procedure, for more details see [22]. The subject sits in front of a computer. A recording consists of multiple trials. Each trial starts with a blank screen. At t=2s, a beep and a cross "+"inform the subject to pay attention. At t=3s an arrow pointing to the left, right, up or down is shown for 1s and the subject is asked to imagine a left hand, right hand, tongue or foot movement, respectively, until the cross disappears at t=7s. This is followed by a 2s break, and then the next trial begins. The EEG data was recorded using 60 electrodes, at a sam-pling rate of 250 Hz and filtered between 1 and 50 Hz. Two independent data files were made available for each subject: training and test.

We applied the same data preprocessing as in our previous work [10] where we re-ported the CFS results. Firstly, we used the Common Spatial Patterns (CSP) method, extended to multiclass problems [23]. It transforms the original signal into a new space where the variance of one of the classes is maximised while the variance of the others is minimized. The result, for each class versus the others, is a new set of 60 signals, ordered based on how informative they are. We selected the first 5 projections and applied 3 frequency band filters (8-12, 12-20 and 20-30 Hz). We then extracted 7 features: max, min and mean voltage values, voltage range, number of samples above zero volts, zero voltage crossing rate and average signal power. This resulted in 420 (5x4x3x7) discrete numeric features.

Table 1 shows the resulting number of instances in the training and test sets. For each subject, the size of the training and test sets were the same. The four classes were equally distributed in both the training and test set, e.g. for subject K3b there were 45 instances from each class in both the training and test data.

Table 1. Number of instances in the training and test sets for each subject

	K3b	K6b	L1b
Training set	180	120	120
Test set	180	120	120

From a data mining point of view the task can be formulated as follows. Given is a training set of 120 or 180 instances, each instance has a dimensionality of 420 fea-tures and is labelled with one of the four classes; the goal is to build a classifier for each subject able to distinguish between the four classes. The curse of dimensionality problem is evident – there are many features but a small number of training instances. It is generally accepted that the number of training instances per class should be at lest 10 times more than the number of features and that more complex classifiers require a larger ratio of sample size to features [7].

3.2   Classification Algorithms

The selected feature sets were tested with 10 classification algorithms which are listed in Table 2. We chose these algorithms as they are state-of-the-art in data mining and also represent different paradigms (rule-based, tree-based, nearest neighbor, probabilistic, function-based, ensemble of classifiers).

It is important to note that the test data was not used in any way during the feature selection. The feature selection was done based on the training data only. A classifier was build using the training data and selected features. It was evaluated on the test data, which was filtered to retain the selected features only.

We used the Weka's implementations [13] of both the feature selection methods and classification algorithms.

Table 2. Classification algorithms used – description and parameters

1R: A rule based on the values of one attribute [19].

Decision Tree (DT): A classical divide and conquer learning algorithm. We used J48.

K-Nearest Neighbor (k-NN): A classical instance-based algorithm; uses normalised Euclidean distance. We used k=5.

Naïve Bayes (NB): A standard probabilistic classifier.

Radial-bases Network (RBF): A two-layer neural network. Uses Gausssians as basis functions in the first layer (number and centers set by the k-means algorithm) and a linear second layer.

Support Vector Machine (SVM): Finds the maximum margin hyperplane between two classes. We used Weka's SMO with polynomial kernel.

Logistic Regression (LogR): Standard linear regression.

Ada Boost (AdaB): An ensemble of classifiers. It produces a series of classifiers iteratively where new classifiers focus on the instances which were misclassified by the previous classifiers and uses weighed vote to combine individual decisions. We combined 10 decision trees (J48).

Bagging (Bagg): An ensemble of classifiers. Uses random sampling with replacement to generates training sets for the classifiers; decisions are combined with majority vote. We combined 10 decision trees (J48).

Random Forest (RF): An ensemble of decision trees based bagging and random feature selection. We used t=10 trees.

4   Results and Discussion

4.1   Feature Reduction

Table 3 lists the number of selected features by the five methods for each subject. It shows that all methods were able to select much smaller subsets of features than the original set of 420 features. The range of the feature reduction was between 53.3% (IG) and 98.1% (Consistency) for K3b, 87.3% (IG) and 98.1% (ReliefF) for K6b and 92.1% (IG) and 97.1% (ReliefF) for L1b. Overall Consistency selected the smallest feature set, followed by CFS, ReliefF, 1RR and IG; the feature set produced by

Table 3. Number of features selected. In brackets is the % of original features retained

	K3b	K6b	L1b	Total
IG	196 (46.7%)	53 (12.6%)	33 (7.9%)	282
CFS	56 (13.3%)	19 (4.5%)	15 (3.6%)	90
ReliefF	82 (19.5%)	8 (1.9%)	12 (2.9%)	102
Consistency	8 (1.9%)	14 (3.3%)	13 (3.1%)	35
1RR	87 (20.7%)	22 (5.2%)	17 (4%)	126

Consistency was eight times smaller than the feature set produced by IG. In [5] CFS was found to select the smallest feature sets on the large datasets, retaining 3-22% of the original features, followed by Consistency, ReliefF and IG. A comparison between CFS and Consistency, the two methods that directly produce feature subsets, shows that in our study CFS retained more features than Consistency (3 times more) while in [5] it retained less (2 times less).

This large feature reduction confirms that the BCI data is noisy and highly correlated. It also reduces the effect of the curse of dimensionality: the ratio of the number of training instances per class to the number of features is reduced from 45/420 to 45/82 – 45/8 for K3b, from 30/420 to 30/56 – 30/8 for K3b and from 30/420 to 30/33 - 30/12 for L1b.

4.2 Classification Performance

Tables 4, 5 and 6 show the classification results in terms of accuracy on the test set for the three subjects, without feature selection and using the five feature selection methods and 10 classifiers. The number of features used is shown in brackets, the best accuracy result for each classifier is in bold and the best accuracy result for the subject is in bold underlined.

A comparison between the subjects shows that the accuracy is highest for K3b and lowest for K6b. This is as expected and due to the different amount of BCI training the subjects received [4]: K3 was the most experienced, L1 had little experience and K6 was a beginner.

Table 4. Subject K3b - accuracy on test set [%] for various classification algorithms, without and with feature selection

	1R	DT	5-NN	NB	RBF	SVM	LR	AdaB	Bagg	RF
none (420)	45.00	67.22	87.78	82.22	82.22	90.56	**88.89**	84.44	84.44	86.67
IG (196)	45.00	67.22	**90.56**	85.55	87.22	**_94.44_**	87.78	80.56	82.78	**89.44**
CFS (56)	45.00	75.00	**90.56**	**91.67**	89.44	93.33	78.88	81.11	81.11	86.11
ReliefF (82)	45.00	**80.56**	88.89	88.33	**89.44**	92.78	80.56	**86.67**	**85.00**	86.11
Cons. (8)	**46.67**	80.00	83.33	81.67	82.22	87.22	81.67	83.89	82.22	82.78
1RR (87)	45.00	76.11	87.88	87.78	87.22	93.89	75.56	85.56	84.44	87.22

Table 5. Subject K6b - accuracy on test set [%] for various classification algorithms, without and with feature selection

	1R	DT	5-NN	NB	RBF	SVM	LR	AdaB	Bagg	RF
none (420)	**40.00**	55.84	**56.67**	50.83	50.83	55.00	45.00	55.83	55.83	50.00
IG (53)	**40.00**	54.17	51.57	**56.67**	55.83	**62.50**	41.67	57.50	58.33	55.83
CFS (19)	28.33	55.84	51.67	54.17	56.67	58.33	**54.17**	61.67	**61.67**	55.83
ReliefF (8)	**40.00**	36.67	37.50	45.00	42.50	45.00	48.33	39.17	40.00	47.50
Cons. (14)	28.33	52.50	52.50	53.33	50.00	57.50	50.83	56.67	56.67	49.17
1RR (22)	**40.00**	**56.67**	52.50	55.83	**58.33**	55.00	47.50	**62.50**	56.67	**60.83**

Table 6. Subject L1b - accuracy on test set [%] for various classification algorithms, without and with feature selection

	1R	DT	5-NN	NB	RBF	SVM	LR	AdaB	Bagg	RF
none (420)	**50.00**	58.33	57.50	60.00	56.67	62.50	60.83	67.50	62.50	48.33
IG (33)	**50.00**	68.33	66.67	63.33	69.17	71.67	59.17	75.00	69.17	**72.50**
CFS (15)	**50.00**	**69.17**	70.83	66.67	68.33	70.83	**70.00**	**78.33**	71.67	63.33
ReliefF (12)	**50.00**	68.33	65.83	70.83	**75.00**	74.17	**70.00**	72.50	68.33	58.33
Cons. (13)	**50.00**	61.67	64.17	**71.67**	67.50	70.00	67.50	68.33	67.50	70.83
1RR (17)	**50.00**	67.50	**70.83**	63.33	63.33	67.50	67.50	66.67	67.50	67.50

It is important to compare the classification accuracy with a baseline. As such we can use the distribution of the majority class in the training data, also called ZeroR prediction [13], which was 25% for all subjects (as noted in Section 3.1 there was no majority class – all four classes were equally distributed). Tables 4-6 show that all classifiers, with and without feature selection, outperformed the baseline.

Table 7. Accuracy on test set [%] – comparison with the top three competition submissions as reported in [21].

BCI team	K3b	K6b	L1b
Hill & Schröder (resampling 100Hz, detrending, Informax ICA, Welch amplitude spectra, PCA, SVM)	96.11	55.83	64.17
Guan, Zhang & Li (Fisher ratios of channel-frequency-time bins, feature extraction, mu and beta band, CSP, SVM)	86.67	81.67	85.00
Gao, Wu & Wei (surface Laplacian, 8-30Hz filter, multi-class CSP, SVM+kNN+LDA)	92.78	57.50	78.33
Ours (CSP, 3 frequency bands, 7 features extracted)	94.44 IG+SVM	62.50 IG+SVM 1RR+AdaB	78.33 CFS+AdaB

Table 7 shows our best results and the results of the top three BCI competition submissions as reported in [21]. Our best results were achieved using feature selection, in particular: IG+SVM for K3b, IG+SVM and 1RR+AdaB for K6b, and CFS+AdaB for L1b. Our results are the second best for each subject, hence they are comparable with the best submitted results.

To rank the feature selection methods, we compared pairwise the accuracy of each two of them, for all subjects. We calculated the number of wins (#wins), draws (#draws) and losses (#losses), and computed the following ranking function: #wins + #draws - #losses. The results are shown in Table 8 for each classifier individually. The last column shows the total score for each feature selector for all classifiers; the higher the score, the better the feature selector. It can be seen that CFS was the best feature selector, followed by IG, 1RR, ReliefF, Consistency and no feature selection. All feature selectors improved the classification accuracy in comparison to no feature selection despite the fact that they discarded a large number of features. CFS, IG and 1RR significantly outperformed the other methods in terms of the wins-draws-losses criterion. From these top three methods, CFS selected the smallest feature set.

Table 8. #wins+#draws-#losses for each feature selector for all subjects

	1R	DT	5-NN	NB	RBF	SVM	LR	AdaB	Bagg	RF	total
none	13	-5	-3	-11	-9	-9	-1	-5	-5	-5	-40
IG	13	-1	5	5	3	13	-7	-1	5	13	48
CFS	5	7	5	7	9	7	7	5	5	1	58
ReliefF	13	3	-5	1	5	-1	3	1	1	-9	12
Cons.	7	-3	-5	-1	-7	-5	3	-3	-3	-5	-22
1RR	13	5	6	5	3	-1	-3	3	3	9	43

Fig.1 shows the effectiveness of the feature selectors when used with various classifiers. For the case without feature selection the best performing classifiers were SVM, AdaB and Bagg; IG performed best with SVM, RF and 5-NN; CFS - with Bagg, AdaB and SVM; ReliefF - with SVM, RBF and NB; Consistency – with SVM, NB and AdaB and 1RR – with SVM, RF and AdaB. As expected, different feature selectors work best with different classifiers [8].

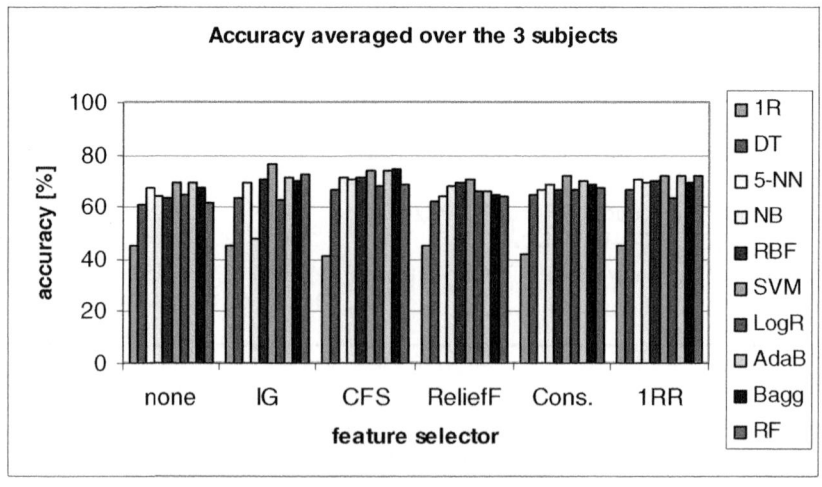

Fig. 1. Comparing feature selectors - accuracy on test set [%] averaged over the 3 subjects

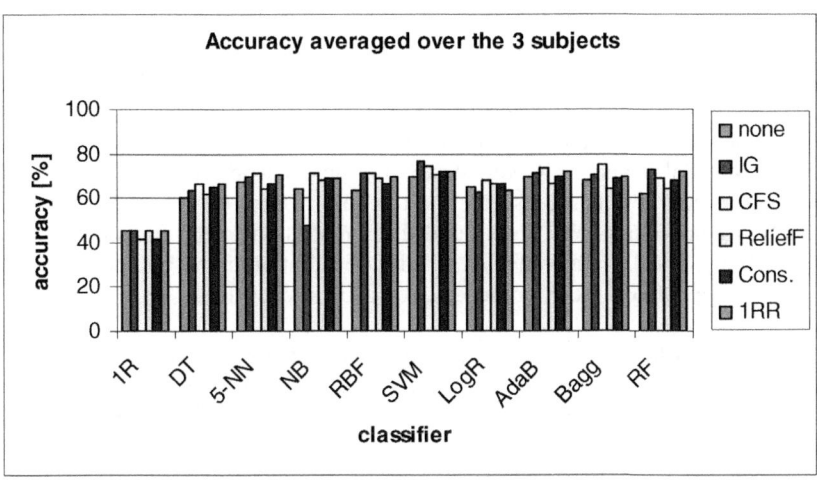

Fig. 2. Comparing classifiers - accuracy on test set [%] averaged over the 3 subjects

A comparison across the classifiers is shown in Fig. 2. The best classifiers were SVM, AdaB, Bagg, RBF and RF. Hence, our results confirm the good performance of SVM from [22] but also show that ensembles of classifiers and RBF are powerful algorithms for classification of BCI data, without feature selection and with the feature selectors we applied. AdaB, Bagg, RF and RBF have received very little attention in previous work on BCI data classification [10]. The widely used linear regression method did not perform well – it ranked 7 out of 10 in terms of accuracy, before 1R, DT and NB. It is also worth noting the poor performance of IG with NB.

SVM was the slowest classifier to build (1.03s to build a classifier for K3B using CFS) as the four-class problem is decomposed into four binary problems, followed by AdaBoost (0.42s) and the remaining three classifiers (0.04-0.19s). In the current BCI systems classifiers are built off-line which means that accuracy is more important than training time; they require fast classification of new data which is true for all except lazy classifiers such as k-NN. However, the need to incrementally retrain the classifier to adapt to the incoming data or subject is recognised as one of the desirable features of the future BCI applications, in which case the training time is important.

5 Conclusions

In this paper we empirically compared five state-of-the-art feature selection methods for classification of BCI data: IG, CFS, ReliefF, Consistency and 1RR. Only CFS has been previously applied to BCI data. We tested the selected feature sets with ten classification algorithms, representing different paradigms, and using benchmark dataset from the BCI competition III. Feature selection was found to be beneficial. In particular, all feature selection methods were found to significantly reduce the number of features (reduction from 53.3% to 98.1%) and to improve accuracy in comparison to the case without feature selection, when used with suitable classification algorithms. Overall, the best feature selector was CFS. In terms of pair-wise accuracy comparison (wins-draws-losses ranking) it achieved the best results followed by IG and 1RR, and it also selected the smallest number of features among the three feature selectors. CFS performed best with ensembles of classifiers such as Bagg and AdaB, and also with SVM.

The best accuracy results per subject were produced by IG with SVM for subject K3b, IG with SVM and 1RR with AdaB for subject K6b, and CFS with AdaB for subject L1b; these results rank second best in comparison to the top results submitted to the BCI competition. We also found that in addition to the popular SVM, other classification algorithms that have received little attention in the BCI community such as AdaB, Bagg, RF and RBF, produced good accuracy results.

Acknowledgments. The data preprocessing was implemented and conducted by Benjamin Harding and Jorge Villalon, and was supported in part by the University of Sydney bridging support grant U1189. This paper is an extended version of [24].

References

1. Dornhege, G., Millán, J.d.R., Hinterberger, T., McFarland, D.J., Müller, K.-R.: Toward Brain-Computer Interfacing. MIT Press, Cambridge (2007)
2. Wolpaw, J.R., Birbaumer, N., McFarland, D.J., Pfurtscheller, G., Vaughan, T.M.: Brain-computer Interfaces for Communication and Control. Clinical Neurophysiology 113, 767–791 (2002)
3. Lotte, F., Congedo, M., Lecuyer, A., Lamarche, F., Arnalsi, B.: A Review of Classification Algorithms for EEG-Based Brain-Computer Interfaces. Journal of Neural Engineering, R1–R13 (2007)
4. Lee, F., Scherer, R., Leeb, R., Neuper, C., Bischof, H., Pfurtscheller, G.: A Comparative Analysis of Multi-class EEG Classification for Brain-computer Interface. In: 10th Computer Vision Winter Workshop (CVWW). Technical University of Graz, Austria (2005)
5. Hall, M.: Correlation-based Feature Selection for Discrete and Numeric Class Machine Learning. In: 17th Int. Conf. on Machine Learning (ICML), pp. 359–366. Morgan Kaufmann, San Francisco (2000)
6. Dash, M., Liu, H.: Feature Selection for Classification. Intelligent Data Analysis 1, 131–156 (1997)
7. Jain, A.K., Duin, R.P.W., Mao, J.: Statistical Pattern Recognition: A Review. IEEE Transactions on Pattern Analysis and Machine Intelligence 22, 4–37 (2000)
8. Hall, M., Holmes, G.: Benchmarking Attribute Selection Techniques for Discrete Class Data Mining. IEEE Transactions on Knowledge and Data Engineering 15, 1437–1447 (2003)
9. Müller, K.-R., Krauledar, M., Dornhege, G., Curio, G., Blankertz, B.: Machine Learning Techniques for Brain-Computer Interfaces. Biomedical Technology, 11–22 (2004)
10. AlZoubi, O., Koprinska, I., Calvo, R.: Classification of Brain-Computer Interface Data. In: 7th Australasian Data Mining Conference (AusDM), pp. 123–132. ACS, Adelaide (2008)
11. Blankertz, B., Curio, G., Müller, K.-R.: Classifying Single Trial EEG: Towards Brain-Computer Interfacing. In: Diettrich, T.G., Becker, S., Ghahramani, Z. (eds.) Advances in Neural Information Processing Systems, vol. 14, pp. 157–164 (2001)
12. Kubat, M., Koprinska, I., Pfurtcheller, G.: Learning to Classify Biomedical Signals. In: Michalski, R.S., Kubat, M., Bratko, I. (eds.) Machine Learning and Data Mining: Methods and Applications. Wiley, Chichester (1998)

13. Witten, I.H., Frank, E.: Data Mining: Practical Machine Learning Tools and Techniques. Morgan Kaufmann, San Francisco (2005)
14. Yang, Y., Pedersen, J.O.: A Comparative Study on Feature Selection in Text Categorization. In: International Conference on Machine Learning (ICML), pp. 412–420 (1997)
15. Kira, K., Rendell, L.: A Practical Approach to Feature Selection. In: 9th Int. Conference on Machine Learning (ICML), pp. 249–256 (1992)
16. Kononenko, I.: Estimating Attributes: Analysis and Extensions of Relief. In: 7th Int. Conference on Machine Learning (ICML), pp. 171–182 (1994)
17. Liu, H., Setiono, R.: A Probabilistic Approach to Feature Selection: a Filter Solution. In: 13th Int. Conference on Machine Learning (ICML), pp. 319–327 (1996)
18. Holmes, G., Nevill-Manning, C.G.: Fearure Selection via the Discovery of Simple Classification Rules. In: Intern. Symposium on Intelligent Data Analysis (IDA), Germany (1995)
19. Holte, R.C.: Very Simple Classification Rules Perform Well on Most Commonly Used Datasets. Machine Learning 11, 63–90 (1993)
20. Cunningham, S.J., Summers, B.: Applying Machine Learning to Subject Classification and Subject Description for Information Retrieval. In: 2nd New Zealand Conference on Artificial Neural Networks and Expert Systems (ANNES), pp. 243–246. IEEE Computer Society, Los Alamitos (1995)
21. Blankertz, B., Müller, K.-R., Krusienski, D., Schalk, G., Wolpaw, J.R., Schlögl, A., Pfurtscheller, G., Millan, J., Schröder, M., Birbaumer, N.: The BCI2000 Competition III: Validating Alternative Approaches to Actual BCI Problems. IEEE Transactions on Neural Systems and Rehabilitation Engineering (2006)
22. Schlögl, A., Lee, F., Bischof, H., Pfurtscheller, G.: Characterization of Four-class Motor Imagery EEG Data for the BCI Competition 2005. Journal of Neural Eng., L14–L22 (2005)
23. Müller-Gerking, J., Pfurtscheler, G., Flyvbjerg, H.: Designing Optimal Spatial Filters for Single-trial EEG Classification in a Movement Task. Clinical Neurophys 110, 787–798 (1999)
24. Koprinska, I.: Comparison of Feature Selection Methods for Classification of Brain-Computer Interface Data. In: Workshop on Advances and Issues in Biomedical Data Mining at the 13th Pacific-Asia Conference on Knowledge Discovery and Data Mining (PAKDD), Bangkok, Thailand, pp. 41–50 (2009)

Mining Protein Interactions from Text Using Convolution Kernels

Ramanathan Narayanan[1], Sanchit Misra[1], Simon Lin[2], and Alok Choudhary[1]

[1] Department of Electrical Engineering and Computer Science,
Northwestern University
[2] Feinberg School of Medicine, Northwestern University

Abstract. As the sizes of biomedical literature databases increase, there is an urgent need to develop intelligent systems that automatically discover Protein-Protein interactions from text. Despite resource-intensive efforts to create manually curated interaction databases, the sheer volume of biological literature databases makes it impossible to achieve significant coverage. In this paper, we describe a scalable hierarchical Support Vector Machine(SVM) based framework to efficiently mine protein interactions with high precision. In addition, we describe a convolution tree-vector kernel based on syntactic similarity of natural language text to further enhance the mining process. By using the inherent syntactic similarity of interaction phrases as a kernel method, we are able to significantly improve the classification quality. Our hierarchical framework allows us to reduce the search space dramatically with each stage, while sustaining a high level of accuracy. We test our framework on a corpus of over 10000 manually annotated phrases gathered from various sources. The convolution kernel technique identifies sentences describing interactions with a precision of 95% and a recall of 92%, yielding significant improvements over previous machine learning techniques.

1 Introduction

Protein-Protein interactions are the associations of protein molecules with one another, and the study of these associations from the perspective of biochemistry, signal transduction and protein networks. Protein interactions form the basis for virtually every process in a living cell, and the study of interactions improves the understanding of diseases and provides researchers with the ability to analyze and study therapeutic approaches.

Over the past few years, a multitude of high throughput methods to detect protein interactions have been developed. Researchers all over the world are performing these experiments and reporting results on protein-protein interactions. Although a few large-scale studies are available, most of the protein-protein interactions come from thousands of smaller studies. Figure 1 shows that the number of articles in Pubmed over the last 50 years is steadily increasing. Consolidating the known list of protein-protein interactions will provide researchers with a powerful tool that will greatly enhance their understanding of these relationships on a genomic scale.

T. Theeramunkong et al. (Eds.): PAKDD Workshops 2009, LNAI 5669, pp. 118–129, 2010.
© Springer-Verlag Berlin Heidelberg 2010

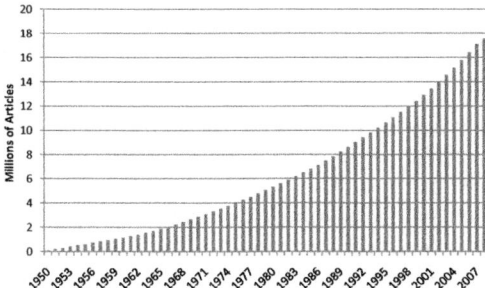

Fig. 1. Growth of the Pubmed Database over the last few years

There have been several attempts to develop databases of interacting proteins, and the supporting metadata that describes an interaction. Currently, there are several manually curated databases like DIP, BIND and MINT [10,1,4]. While these databases promise a high degree of accuracy, by providing supporting evidence reviewed by a subject expert, their coverage leaves much to be desired. DIP contains 56493 records referring to protein interactions whereas MINT describes 105537 interactions. However, there is a common consensus that the number of interactions available in these manually curated databases is miniscule compared to the total number of interactions described in the literature. Typically, manually curated databases cover about 3000-4000 articles. Considering that Pubmed, the biomedical literature database maintained by National Institute of Health (NIH) contains over 17 million documents, we can clearly see the limited usefulness of manually curated databases.

Machine learning and data mining techniques have been applied to develop automated and semi-automated methods for finding protein interactions on a large scale. The highly ambiguous nature of natural languages and lack of standards in research writing complicate this task tremendously. The lack of standards in protein nomenclature has led to a single protein being referred to by several synonyms, with no apparent similarities (eg. Trehalose-phosphatase, EC 3.1.3.12 and TPP refer to the same protein). Therefore, the use of standard exhaustive dictionaries to extract protein/gene references [2,17] have not been very effective. Existing machine learning approaches [15,14] have shown some success, but suffer from lack of scalability, inability to adapt to different domains and small test sets. Semi-automated techniques like PreBIND use SVMs to speed up manual expert reviewing by reducing the amount of information to be examined. The challenges of manual curation can be addressed by developing fully automated systems to extract interactions from abstracts or full text articles. However, fully automated methods suffer from low sensitivity and a lack of coverage. Since there are a large number of ways in which interactions can be described, simple rule-based approaches, relying on human-generated rules to recognize phrases, can only capture a limited percentage of interactions. At the same time, using a complex but accurate machine learning technique may not be computationally feasible, nor achieve the desired coverage. Previous efforts have also been hampered by the lack of quality training datasets.

However, we utilize the highly accurate protein interaction information in interaction databases like DIP and MINT to serve as a vast, diverse training platform for our supervised learning system. In this paper, our contributions are

- We propose a scalable hierarchical Support Vector Machine(SVM) based framework to efficiently mine protein interactions from Pubmed abstracts with high precision.
- We propose the use of Convolutions kernels in Support Vector Machines for mining protein-protein interactions.
- We validate our technique on a large corpus gathered from various manually annotated databases, and achieve high rates of precision and recall.

2   Kernel Methods and Support Vector Machines

Support vector machines (SVMs) are a set of supervised machine learning techniques used for classification and regression. SVMs have been shown to be highly successful for text classification[11]. [19] contains a comprehensive discussion of support vector machine theory. SVMs are able to solve a multitude of classification problems by using domain-specific and cost-sensitive kernel functions. In our work, we use the concept of tree kernels to efficiently determine the similarity between the syntactic parse trees of sentences. As a result, we are able to exploit the syntactic structure of natural language text and develop a better classification model when compared to a traditional SVM.

3   Hierarchical Approach

We tackle the problem of finding protein interactions in Pubmed abstracts using a three-stage hierarchical approach. Figure 2 shows a brief overview of our approach, and the various stages involved.

Stage I: Pubmed contains a vast number of abstracts (around 17 million), most of which do not contain descriptions of protein interactions. Comprehensive analysis of each of these abstracts is time-consuming and wasteful in terms of computational resources. Therefore we need a way to determine which documents are worth looking at in greater detail. This is a classic example of a text classification problem, and we propose a solution using a simple SVM formulation.

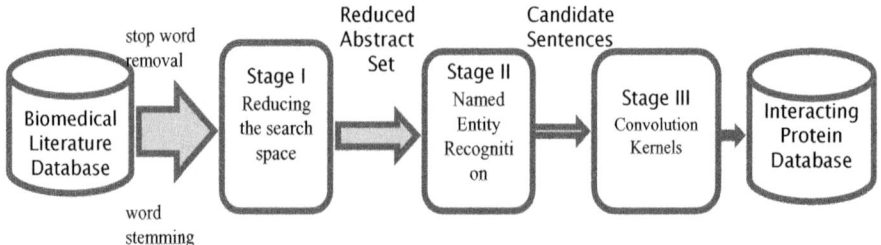

Fig. 2. System architecture for Mining Protein-Protein Interactions

Stage II: Once we have determined those abstracts which are likely to contain interactions, we need to narrow down the list of possibilities furthur. Occurrence of proteins (in general, biomedical named entities) is a strong requirement for description of an interaction. In other words, it is very likely that a sentence containing two or more protein names actually describes an interaction amongst them. Also a sentence containing no references to proteins in unlikely to contain an interaction. To find proteins in text, we use a comprehensive SVM-based method, that yields high recall, thus finding most protein references, and by implication, most sentences that describe interactions.

Stage III: Finally, to determine if a sentence describes an interaction or not, we must understand the inherent meaning of the sentence. This can be a tricky proposition, considering the richness and diversity of the English language. However, by studying the syntactic and semantic structure of the text, we can determine, with high accuracy, whether or not it describes an interaction. For this purpose, we introduce the concept of convolution kernels to accurately identify sentences describing interactions. While this is a resource intensive technique, we only apply it to those candidate sentences which have been obtained after the first two stages.

3.1 Reducing the Search Space

In our problem, we require a method to prune the search space, so that we can apply more accurate and complex machine learning techniques on a reduced candidate set. We have used SVMs with a ***Bag-of-Words*** approach to build a classifier model which successfully distinguishes those abstracts which may contain a protein interaction. The Bag-of-words technique is a simple approach that relies on word frequency information to classify text documents. Each abstract is converted into a high dimensional feature vector, after performing necessary pre-processing steps like stemming and stop-word removal. We used the TFIDF [11] metric to represent features in the input vector. The details of the training and validation phases are provided in the results section. The end-result of this phase is that we are able to identify those documents which are likely candidates for containing protein-protein interactions, with an acceptable false-positive rate.

3.2 Biomedical Named Entity Recognition (NER)

Stage II involves the recognition of biomedical named entities(more specifically, proteins) in text. In this work, we do not distinguish between proteins and other named entities(NEs), since it is difficult to make such a distinction. Each word is represented by a binary feature vector, which consists of lexical features, orthographical and morphological features, Part-of-Speech features, and context features(as in [13]). The basic idea is that using these features, the SVM will be able to determine whether the word is a biomedical named entity. The features we use are similar, but not identical to those in [8]. Details of these features are discussed below:

Lexical Features: The lexical feature set of a word consist of three lists of terms: single-term list, functional-term list and general-term list. Each term in these lists corresponds to a feature. Single-terms are words which can be used as an protein entity by themselves, such as 'NF-KappaB', 'astrogen' or 'ERK'. Functional terms are devised to describe the function and characteristics of NEs. They have no special orthographic features and consist of all lower case words which frequently appear in NEs (eg. 'protein','gene','receptor' are all functional words). General terms are a set of words that are classified neither as single-terms nor function terms. This is simply a list of words with frequency greater than three in the training corpus.

Orthological and Morphological Features: A large number of NEs contain surface clues (called orthological features), which may help in discriminating them. We use a list of sufficiently powerful orthological features in our model(see Figure 3). We also utilize commonly occurring suffix patterns in NEs as a predicting feature.

Part-of-Speech(POS) Features: The part-of-speech tags of a target word and its surrounding context words represent syntactic characteristics for composing an entity. It is commonly seen that named entities are tagged as nouns or adjectives, whereas they are rarely tagged as adverbs.

Contextual features: For boundary identification, we use neighboring words and the POS of neighboring words as contextual features. In our experiments, we considered the two words to the left and right of the target word. Context words are selected as features only if they belong to one of the lexical feature lists.

Therefore, the feature vector for a target word is obtained by composing all of the features described above. For each feature, the binary feature vector has a 1 if that feature is present, or 0 otherwise. We use a SVM with standard settings and a linear kernel. Once biomedical entities are recognized, we will be able to further narrow down the list of candidate sentences which may contain protein interactions.

3.3   Convolution Kernels

Stage III represents the most complex of all our SVM formulations. The input is a sentence containing several biomedical named entities. However, the mere presence of a protein does not guarantee that the sentence will describe a protein interaction (See Figure 5). Also the use of the Bag-of-Words and similar techniques does not allow us to exploit the syntactic structure of the phrase in question. In recent years, tree kernels have been shown to be interesting approaches for the modeling of syntactic information in natural language tasks [21,22,23].

Given an input sentence, we can obtain the corresponding syntactic parse tree using standard natural language processing techniques. Figure 4 shows the parse tree of the sentence 'IL-5 activated the Jak 2-STAT 1 signaling pathway'. The kernels that we consider represent parse trees in terms of their substructures(fragments). These substructures define the feature space of a tree, which is represented as a high-dimensional vector. The kernel function attempts to find the similarity between two trees by counting the number of their common fragments. When this kernel function is plugged into a SVM, it detects if a parse tree belongs to the feature space of the known examples belonging to the target class, thereby accomplishing the classification task.

Feature ID	Example
Cap + Digit	IL22
Alpha-numeric	Ama292
Cap + Digit + Cap	L809TR
Greek	NF-KappaB
Lower+digit+sym	110-nt
Unit	U/mg
HasDot	a.m
Has_special_symbol	Doc1/Apc10

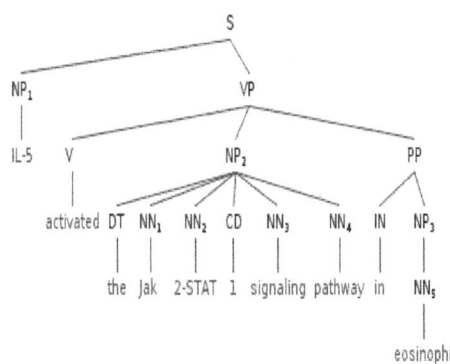

Fig. 3. An example of Orthological Features used for NER

Fig. 4. Parse Tree

In a syntactic parse tree, each node with its children is associated with a grammar production rule, where the symbol on the left-hand side corresponds to the parent node, and the symbols on the right hand side are associated with the child nodes. We define as a subtree(ST) any node of a tree along with all its descendants. A subset tree (SST) is a tree obtained by applying the same grammatical rule set which generated the original tree. The main idea behind tree kernels is to compute the number of common substructures between two trees T_1 and T_2 without explicitly considering the whole fragment space. We base our tree kernel on the method proposed in [5], along with a modification that enables us to evaluate differences between features spaces generated from STs or SSTs.

We formally represent a parse tree feature space (ST or SST) as $T = (f_1, f_2, \dots f_{|\mathcal{N}|})$, where f_i represents a ST or SST fragment. Define a function $F_i(n)$ to be 1 if fragment f_i is rooted at node n of T, and 0 otherwise. The kernel function K between two trees T_1, T_2 is defined as follows: $K(T_1, T_2) = \sum_{n_1 \in N_{T_1}} \sum_{n_2 \in N_{T_2}} \sum_{i=1}^{|\mathcal{N}|} F_i(n_1) F_i(n_2)$.

Here N_{T_1}, N_{T_2} represent node sets of T_1, T_2, and $|\mathcal{N}|$ represents the size of the feature space.

It can be shown that computing the kernel function as-it-is can lead to an exponential number of evaluations in the size of the input tree. However, we can compute the inner product ($\sum_{i=1}^{|\mathcal{N}|} F_i(n_1) F_i(n_2)$, designated as $IP(n_1, n_2)$) in an efficient manner as follows:

- If the productions at n_1 and n_2 are different, then $IP(n_1, n_2) = 0$
- If the productions at n_1 and n_2 are the same, and n_1 and n_2 have only leaf children, then $IP(n_1, n_2) = \lambda$
- If the productions at n_1 and n_2 are the same, and n_1, n_2 are not pre-terminals, then

$$IP(n_1, n_2) = \lambda \prod_{j=1}^{numchild(n_1)} (\sigma + IP(c_{n_1}^j, c_{n_2}^j))$$

where c_n^j is the jth child of node n. When $\sigma = 0$, we evaluate the ST kernel, and when $\sigma = 1$, the SST kernel is evaluated. λ is a decay parameter that determines the importance of the tree fragment length in the kernel evaluation.

The complexity of evaluating the above kernel function is quadratic in the length of the trees. However using several optimization techniques, the kernel evaluation can run in linear time on average. Therefore, in Stage III we generate the Part-of-Speech tags for each sentence and then generate the corresponding syntactic parse tree. In addition to the parse tree representation, we also generate a bag-of-words feature vector for each candidate sentence. This is because we would like to combine the word-frequency information with the syntactic tree kernel to generate a hybrid kernel. The details of the training and test datasets, implementation and results are discussed in the next section.

4   Experimental Results

In order to verify the effectiveness of our system, we perform extensive testing to gauge the performance of all three stages individually and as a group. We use SVM-light [12] software for all our experiments. The metrics we use to determine performance are Precision, Recall and F-1 score.

Stage I: We generated a training set of 3730 Pubmed abstracts (2230 positive cases and 1500 negative cases). The positive examples were those abstracts in DIP(Database of Interacting Proteins) which have been manually curated and found to contain protein interactions. The negative examples were derived by selecting a subset of Pubmed abstracts which had no interaction keywords, and then were reviewed by a subject expert. After performing preprocessing steps like word stemming and stop word removal, the SVM was trained with these examples using a linear kernel and default settings. Testing the model on a set of 500 positive examples and 1000 negative examples(generated from the same sources as described above) yielded a precision of 97.5% and recall of 92%. For further testing, we obtained a set of 3121 positive examples from another manually curated database, MINT, and attempted to classify those abstracts using our SVM model. The model was able to successfully predict that 96.4% of these abstracts described protein interactions. Therefore, with a lightweight technique, we are able to dramatically reduce the search space, without generating too many false negatives.

Stage II: The Genia corpus [9] is a large human subject expert annotated database which serves as an ideal platform to train and test supervised learning systems. In the Genia corpus, 2000 abstracts from Pubmed have been manually annotated to denote biological entities like proteins, genes, DNA etc. as well as language constructs like sentences, abbreviations and titles. The number of 'protein' terms alone is 10504, which indicates the richness of this corpus.This serves as an ideal testing platform to measure the effectiveness of the Named entity recognition module. We used the GENIA POS tagger [9] for part-of-speech tagging, and computed the various lexicographic and orthographic features for each word as described in the earlier section. Inspite of inconsistent naming conventions and punctuation, our named entity recognition module was able to

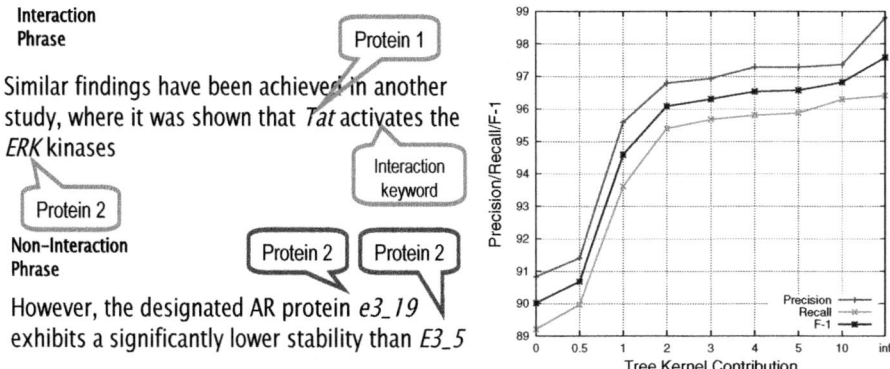

Fig. 5. Positive and Negative examples from the Input Corpus

Fig. 6. Performance comparison of the Hybrid kernel for different values of τ

successfully identify biomedical named entities (NEs) with a precision of 79% and a recall of 77%.

Stage III: In order to generate an appropriate dataset to test the convolution kernel SVM formulation, we consider a list of protein interaction sentences generated by the BioText project [16]. This dataset contains a list of 3190 sentences which describe protein interactions of various kinds. A list of 7000 negative examples was generated in a semi-automated fashion by gathering sentences which contained protein references, yet did not contain interaction keywords. Figure 5 shows examples of positive and negative cases in the corpus. The sentences were divided into training and test datasets. The Genia POS tagger was used for part-of-speech tagging, and the Collins parser [6] was used to generate syntactic parse trees.

In our experiments, we first compare the performance of the convolution kernel based method with a Bag-of-words approach. Further, we use a combination of both these approaches to generate a hybrid model. The kernel formulation for two input examples $x_1{:}(T_1, v_1)$, $x_2{:}(T_2, v_2)$ where T_i represents the syntactic parse tree, and v_i represents the traditional bag-of-words feature vector is as follows:

$$K(x_1, x_2) = \tau K_t(T_1, T_2) + K_{bow}(v_1, v_2),$$

where τ is a parameter that represents the contribution of the tree kernel, K_t represents the tree kernel, and K_{bow} represents a traditional SVM kernel on a bag-of-words feature vector(radial,gaussian etc). Figure 6 shows the variation in precision and recall as we vary the contribution of the tree kernel (parameter σ is set to 0). It can be seen that the usage of a tree kernel significantly improves the precision and recall values. This is because the tree kernel takes into account the syntactic structure of a sentence, and is able to capture the inherent semantics in a superior way, as compared to a normal bag-of-words feature representation. In fact, the use of word frequency information may actually bias the classifier and produce erroneous results. The highest precision and recall values are when $\tau = \infty$, or in other words, when only the tree kernel is considered.

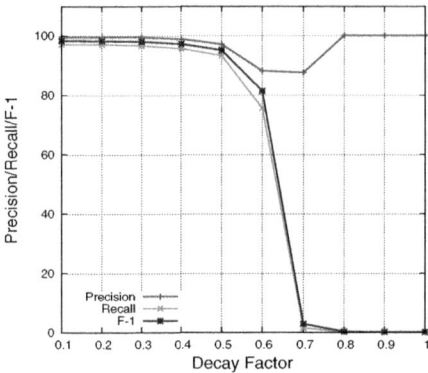
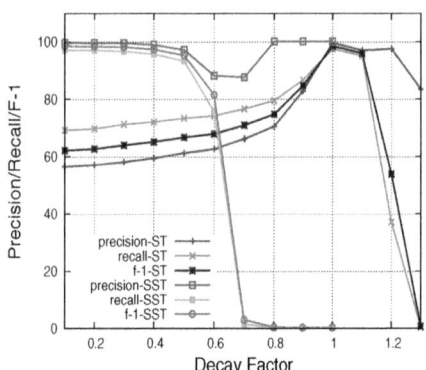

Fig. 7. Performance comparison of the Tree kernel for different decay factors

Fig. 8. Performance comparison of Subtree Kernel vs. SST Kernel

We also study how our classification model changes when we change the value of λ, the tree decay parameter. Intuitively, λ adjusts the importance of tree fragments according to their size. A higher value of λ indicates that we are concerned about the details of the fragment structure, especially for larger sentences. Figure 7 shows how the precision, recall and F-1 values vary for λ in the range $(0,1]$. The performance of the kernel remains relatively stable for λ in the range $[0.1-0.5]$, with a dramatic drop around 0.6. At this stage, the precision values reach 100%, but the recall falls to near-zero levels. This indicates that our SVM based model performs better when we consider the local structure of the fragments, without giving too much weight to the details. The ideal choice of λ is determined to be 0.4.

We also ran experiments testing the performance of the subtree kernel as compared to the subset tree kernel. The difference between these kernels is the value of σ. For the subtree Kernel, $\sigma = 0$, and for the subset tree kernel $\sigma = 1$. The results described earlier are for the subset tree kernel. Figure 8 shows the comparative performance between the subtree and subset tree kernels for different values of the decay parameter. The subtree kernel performance for lower values of λ is poor, but it gradually improves till at $\lambda = 1$, the performance matches the best case performance of the subset tree kernel. The optimum choice of λ for the subtree kernel is 1. Overall, we were able to achieve an average precision of 95% and average recall of 92% using the convolution kernel technique.

Finally, to test the system as a whole, we ran experiments to evaluate the performance on a dataset of 1584 Pubmed abstracts. After Stage I, we were able to identify 95% of those abstracts which did contain an interaction. Similarly, Stage II was able to identify 82% of all protein references. Finally in Stage III we were able to identify 95% of those sentences which contained protein interactions. We also tested our technique on the Biocreative 2[27] dataset. However, the toughest task in this dataset is the Named Entity Recognition, and the precision and recall of the results are highly influenced by this step. Since that is not the major focus of this work, we have evaluated only Stage III of our technique using this dataset. The convolution kernel technique yields a precision of 93% and a recall of 89% on a test set comprising of over 2000 sentences.

Previously used techniques like [18,8,20] make use of manually constructed interaction phrases, as well as context free grammars to describe interaction patterns between biological entities. The interactions generated by these techniques are limited by the coverage of the recognition rules, as well as the inherent complexity and variability in sentences describing interactions. These methods achieve high precision rates (above 90%), but suffer from extremely low recall rates (around 20-30%). Another popular approach is to use machine learning and statistical techniques [15,14], which have been shown to have higher recall. However, the major issues are scalability and portability. State-of-the-art text mining systems for protein interaction mining([3]) which offer full automation are tested on datasets of only a few hundred articles. In contrast, we have validated our results using a much larger corpus. Shin et al[28] also use tree kernels to identify interaction sentences. However, our work differs from theirs in that we explore a variety of different kernels (and combinations of convolution/bow kernels), and perform extensive performance comparisons. Though the convolution kernel technique performs impressively, as a whole, the overall recall of our system is low (around 60%). This is because there is loss of interaction information in each stage.

5   Conclusions and Future Work

In this paper, we have highlighted the need for a scalable, accurate method for predicting protein-protein interactions from text articles. We propose a hierarchical Support Vector Machine based system to efficiently mine these interactions from Pubmed abstracts. In our three-stage system, we use simple word-frequency based SVM formulation in Stage I, a slightly more complex Named Entity Recognition module in Stage II, and a sophisticated, accurate convolution kernel-based method in Stage III. At each stage, the complexity of the technique used increases. However, a large number of redundant documents are eliminated in the earlier stages, thereby reducing the overall workload. In doing so, we achieve a reasonable performance/accuracy tradeoff. Extensive testing on real-world datasets yields reasonable results. In the future, we plan to address the problem of extracting the specific interaction type. Also, while abstracts contain useful information, a vast body of knowledge lies hidden in full text articles. It will be useful to extend our technique to mine interactions from full text articles.

Acknowledgements

This work was supported in part by NSF grants CNS-0551639, IIS-0536994, NSF HECURA CCF-0621443, and NSF SDCI OCI-0724599 and DOE SCIDAC-2: Scientific Data Management Center for Enabling Technologies (CET) grant DE-FC02-07ER25808.

References

1. Alfarano, C., et al.: The Biomolecular Interaction Network Database and related tools 2005 update. Nucleic Acids Res. 33, D418–D424 (2005)
2. Blaschke, C., et al.: Automatic extraction of biological information from scientific text: protein-protein interactions. In: Proc. Int. Conf. Intell. Syst. Mol. Biol., pp. 60–67 (1999)
3. Brown, K.R., et al.: Online predicted human interaction database. Bioinformatics 21, 2076–2082 (2005)
4. Chatr-aryamontri, A., et al.: MINT: the Molecular INTeraction database. Nucleic Acids Res. 35, D572–D574 (2007)
5. Collins, M., Duffy, N.: New ranking algorithms for parsing and tagging: Kernels over discrete structures (2002)
6. Collins, M.: Head-Driven Statistical Models for Natural Language Parsing. Computational Linguistics (2003)
7. Donaldson, I., et al.: PreBIND and Textomy–mining the biomedical literature for proteinprotein interactions using a support vector machine. BMC Bioinformatics 4, 11 (2003)
8. Fukuda, K., et al.: Toward information extraction: identifying protein names from biological papers. In: Pac. Symp. Biocomput., pp. 707–718 (1998)
9. Genia Project: Mining literature for knowledge in molecular biology (2008), http://wwwtsujii.is.s.u-tokyo.ac.jp/GENIA/home/wiki.cgi
10. Gilfillan, I.: A database of proteins that are known to interact. Genome Biology 1; Reports220 (November 2000)
11. Joachims, T.: Text categorization with support vector machines: learning with many relevant features. In: Nédellec, C., Rouveirol, C. (eds.) ECML 1998. LNCS, vol. 1398, pp. 137–142. Springer, Heidelberg (1998)
12. Joachims, T.: Making large-scale SVM learning practical. In: Advances in Kernel Methods-Support Vector Learning (1999)
13. Lee, K.J., Hwang, Y.S., Kim, S., Rim, H.C.: Biomedical named entity recognition using two-phase model based on SVMs. J. Bio. med. Inform. 37, 436–447 (2004)
14. Marcotte, E.M., et al.: Mining literature for protein-protein interactions. Bioinformatics 17, 359–363 (2001)
15. Ramani, A.K., et al.: Consolidating the set of known human protein-protein interactions in preparation for large-scale mapping of the human interactome. Genome Biol. 6, R40 (2005)
16. Rosario, B., Hearst, A.: Multi-way Relation Classification: Application to Protein-Protein Interaction. In: Human Language Technology Conference on Empirical Methods in Natural Language Processing (2005)
17. Rindflesch, T.C., et al.: Mining molecular binding terminology from biomedical text. In: Proc. AMIA Symp., pp. 127–131 (1999)
18. Temkin, J.M., Gilder, M.R.: Extraction of protein interaction information from unstructured text using a context-free grammar. Bioinformatics 19, 2046–2053 (2003)
19. Vapnik, V.N.: The Nature of Statistical Learning Theory. Springer, Heidelberg (1995)
20. Yu, H., et al.: Automatic extraction of gene and protein synonyms from MEDLINE and journal articles. In: Proc. AMIA Symp., pp. 919–923 (2002)
21. Culotta, A., Sorensen, J.: Dependency Tree Kernels for Relation Extraction. In: Proceedings of ACL 2004 (2004)
22. Bunescu, R., Mooney, R.J.: Subsequence kernels for relation extraction. In: Proceedings of the 19th Conference on Neural Information Processing Systems, Vancouver, British Columbia (2005)

23. Collins, M., Duffy, N.: Convolution kernels for natural language. In: NIPS 2001 (2001)
24. Yuka, T., Tsujii, J.: Part-of-Speech Annotation of Biology Research Abstracts. In: The Proceedings of 4th International Conference on Language Resource and Evaluation (LREC 2004), Lisbon, Portugal, May 2004, pp. 1267–1270 (2004)
25. Collins, M.: A New Statistical Parser Based on Bigram Lexical Dependencies. In: Proceedings of the 34th Annual Meeting of the ACL, Santa Cruz
26. Porter, M.F.: An algorithm for suffix stripping. Program 14(3), 130–137 (1980)
27. Biocreative 2: http://biocreative.sourceforge.net/biocreative_2.html
28. Shin, et al.: Identifying Protein-Protein Interaction Sentences Using Boosting and Kernel Method. In: Second BioCreative Challenge Evaluation Workshop (2007)

Missing Phrase Recovering by Combining Forward and Backward Phrase Translation Tables

Peerachet Porkaew and Thepchai Supnithi

Human Language Technology Laboratory
National Electronics and Computer Technology
112 Thailand Science Park, KlongNueng, KlongLuang,
Pathumthani, Thailand 12120
{Peerachet.Porkaew,Thepchai.Supnithi}@nectec.or.th

Abstract. We propose a method to recover missing phrases dropped in the phrase extraction algorithm. Those phrases, therefore, are not translated even though we tested the system with the training data. On the other hand, in native-to-foreign, or backward training, some missing phrases can be recovered. In this paper, we combined two phrase translation tables extracted by the source-to-target and target-to-source training for the sake of more complete phrase translation table. We re-estimated the lexical weights and phrase translation probabilities for each phrase pair. Additional combining weights were applied to both tables. We assessed our method on different combining weights by counting the missing phrases and calculating the BLEU scores and NIST scores. Approximately 7% of missing phrases are recovered and 1.3% of BLEU score is increased.

Keywords: statistical machine translation, phrase translation table, parallel phrase extraction.

1 Introduction

The early SMT system, introduced by IBM [1], is based on word alignment probabilities. To enhance the accuracy,features of the translation model such as fertility and distortion were incorporated to the system. A well-known word alignment toolkit GIZA++ is developed based on IBM models. However, using only word translation probabilities leads the system to the ambiguity of translation alternatives as local contexts are predominantly taken into account. Afterwards, the phrase-based approach, focusing on groups of connecting words, was emerged to resolve this deficiency. This approach claims to yield higher-quality result compared to the word-based approach. The word translation model was adapted to the phrase translation model [2] using phrase translation table. The efficiency of phrase translation table is affected by two main factors, i.e. (1) the correctness of phrase pair and (2) phrase scores such as translation probabilities.

Building phrase table is a knowledge acquisition process for the system. Okuma [3] introduced adding a dictionary into the phrase-based system with reordering information. In other researches, factors such as word form, root of word,

T. Theeramunkong et al. (Eds.): PAKDD Workshops 2009, LNAI 5669, pp. 130–140, 2010.

morphological information and part-of-speech are applied in the language model [4] and the translation model [5].

We organize this paper as follows. An example of missing phrases is illustrated in Section 2. Section 3 describes the methodology for solving "missing phrases". The experiment settings are shown in Section 4. In Section 5, we assess our method and discuss the experiment results. Finally, we conclude our paper and list up future work Section 6.

2 Background

In word-based SMT, the translation is model on probabilities in the word level. The translation result is produced by choosing the best word-to-word translation option. This model cannot cope with the ambiguity caused bymultiple-word translation. Phrase-based translation was then emerged to fulfill this deficiency. Besides word-level translation, the phrase-based model utilizes the probability of chunk-and phrase-level translation in dealing with multiple-word translation. This model outperforms the word-based model because it is capable of capturing local context information. Fig. 1 shows the difference between word-based translation model and phrase-based onewhen translating two phrases "international school" and "international politics". Translating the word "international" in "international school" by using word translation probabilities might encounterthe translation ambiguity,where the word "international" can equally be translated to either "นานาชาติ "or "ระหว่างประเทศ". However, this problem can be solved by using phrase translation probabilities which are estimatedfrom the phrase extraction algorithm.

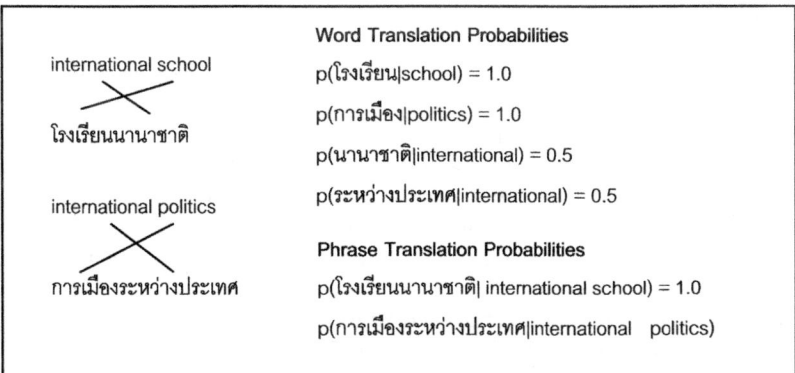

Fig. 1. Word and phrase translation probabilities of "international school" and "international politics"

In subsections 2.1 to 2.3, we review the phrase extraction algorithm and provide an example of the missing phrase problem. Then, we illustrate the overview of our solution.

2.1 Phrase Extraction Algorithm

The heuristic algorithm [2] for the phrase extraction starts with an intersection of bidirectional word alignments. Then, the growing algorithm is applied for adding alignments based on the union of the both directions of word alignments. All consistent phrase pairs are extracted. Fig. 2 shows an inconsistent and a consistent phrase pairs, respectively. As seen, the consistent phrase pair does not contain any alignments outside the scope represented in the rectangle, while the inconsistent one does.

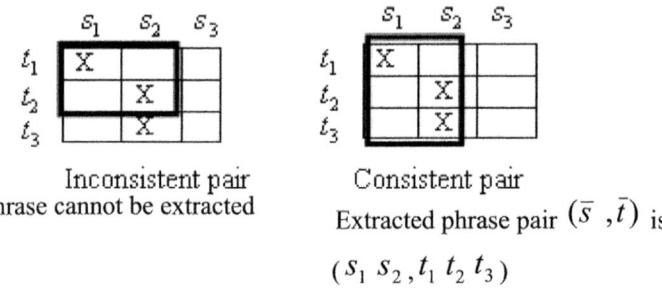

Inconsistent pair
Phrase cannot be extracted

Consistent pair

Extracted phrase pair $(\bar{s}, \bar{t})$ is

$(s_1 \, s_2, t_1 \, t_2 \, t_3)$

Fig. 2. Inconsistent and consistent phrase pair

2.2 Problem

During the development of English-to-Thai SMT, we investigated the translation quality by using the training data for a test set. We found that, surprisingly, some phrases were incorrectly translated. Once we recognized this problem, we then focused on the phrase extraction algorithm.

Normally, if the phrase extraction algorithm cannot find any consistent alignment within a certain number of words (our default is 7 words), that phrase pair will be dropped. In addition, all words in the dropped phrase pairs that appear only once in the corpus will becomeunknown words when we translate them. We call those unknown phrases as missing phrases.

For example, Fig. 3 compares Thai-to-English and English-to-Thai word alignments. Fig. 3(a) shows an example of alignments that any phrase pairs cannot be extractedfrom by two reasons. First,the phrase pair in the inner rectangle is inconsistent because the alignment X* is aligned outside itself. Second,while the phrase pair in the outer rectangle is consistent, the number of target word exceeds the limitation so it is dropped out. Fig. 3(b) shows the transposed word alignment of Thai-to-English. Comparing to the inner rectangle of Fig. 3(a), the rectangle in Fig. 3(b) is consistent andthenumber of words in the source and target sentences donot exceed the limit number.

2.3 Proposed Solution

Intuitively, allowing more words in the rectangle may solve the problem, but the phrase table might over-fit the training corpus. Based on our investigation shown in the example, we propose to add all phrases extracted from the Thai-to-English word alignment into the English-to-Thai phrase translation table.

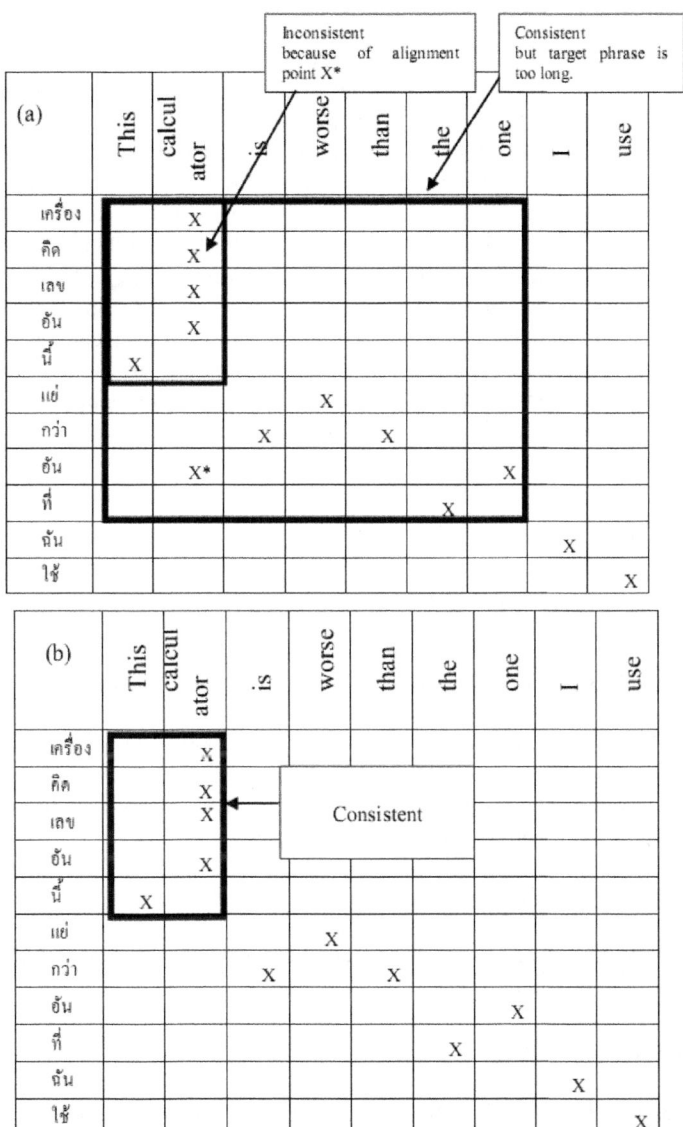

Fig. 3. (a) Word alignment of English-to-Thai; (b) Transposed word alignment of Thai-to English

The phrase extraction algorithm produces a symmetric alignment by using the intersection of bidirectional word alignments. The bidirectional word alignment, probabilities of source word given target word and probabilities of target word given source word, is trained by GIZA++, which is based on EM-algorithm. Assume that we have two languages which are A (source) and B (target), the GIZA++ will generate the word translation probabilities of A given B $P_{forward}(A \mid B)$ and the word translation probabilities of B given A $P_{forward}(B \mid A)$. We call these word translation models as *forward* models because we set A to be the source language and set B to be the target language. These *forward* word translation models are used to produce bidirectional word alignments by choosing the best translation of each word.

However, if we switch the source language and the target language (set A to be target and B to be source) and then we train them with the GIZA++ again, we will get another word translation models, including translation probabilities of B given A $P_{backward}(B \mid A)$ and translation probabilities of A given B $P_{backward}(A \mid B)$. We call these word translation models as *backward* model. By investigation, the *forward* word alignment model and the *backward* word translation model can be different. We found that many correct phrases in *backward* phrase translation table are not included in *forward* phrase translation table. According to this phenomenon, we aim to produce better phrase translation table by combining *forward* and *backward* phrase translation tables.

3 Methodology

In combining process, we defined three processes for calculating phrase translation probabilities and lexical weights.

1. Word level combining process – To combine word translation probabilities that are used for the lexical weight calculation.

2. Phrase level combining process – To combine phrase translation probabilities of the forward phrase translation table and the backward phrase translation table.

3. Lexical weight calculation process – To calculate lexical weight of each phrase pair using combined word translation probabilities.

3.1 Word Level Combining Process

We developed an *adding algorithm* to combine word-based translation model of an English word given a Thai word.

3.2 Adding Algorithm

Definition. Let $\{e_1, e_2, ..., e_I\}$ be the set of English words and $\{t_1, t_2, ..., t_J\}$ be the set of Thai words of a given corpus. We define $w_{forward}(e_i \mid t_j)$ and $w_{backward}(e_i \mid t_j)$ to be the probability of Thai word t_j which is translated into English word e_i in forward and backward model, respectively.

In the *adding algorithm*, we gradually add each translation pair of *backward model* into the original *forward model* and compute a scoring function shown in Eq.4. This scoring function $\tau(e_k, t_i)$ represents how likely that given t_l is translated into the target e_k after determining the forward translation models and backward translation models.

$$\tau(e_k, t_l) = \alpha \times w_{forward}(e_k \mid t_l) + (1 - \alpha) \times w_{backward}(e_k \mid t_l) \qquad (1)$$

For $k = 1, 2, ..., I$ and $l = 1, 2, ..., J$

Where α is a combining weight that balances the effectsofthe forward and backward models.

We normalized $\tau(f_k, e_l)$ to construct   a combined word-based translation model of English words given Thai words by Eq.5.

$$\widehat{w}_{forward}(e_k \mid t_l) = \frac{\tau(e_k, t_l)}{\displaystyle\sum_{m=1}^{I} \tau(e_m, t_l)} \qquad (2)$$

For $k = 1, 2, ..., I$ and $l = 1, 2, ..., J$

We normalized $\tau(e_k, t_l)$ due to the property of conditional probability which is

$$\sum_{m=1}^{I} \widehat{w}(e_m \mid t_l) = 1 \quad , \text{ for } l = 1, 2, ..., J \qquad (3)$$

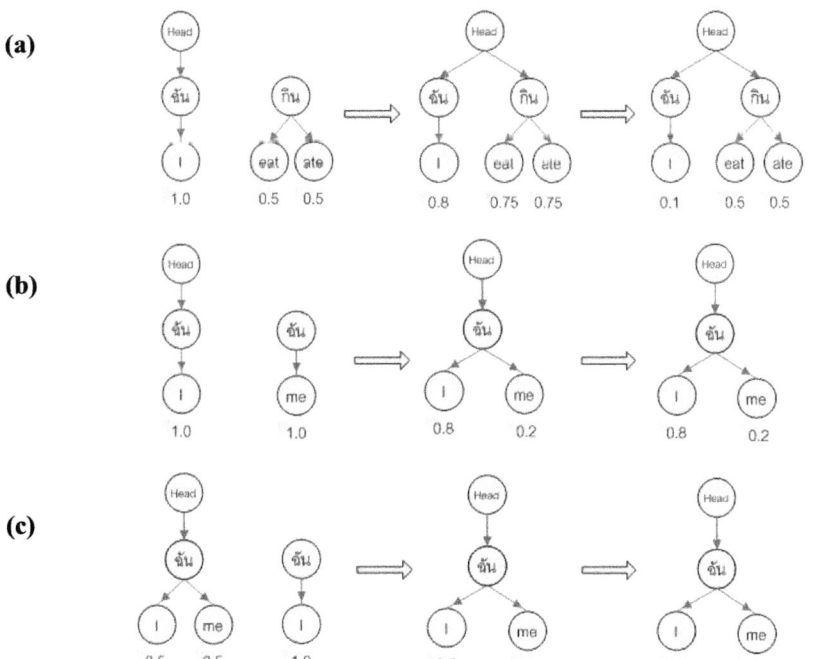

Fig. 4. Graphical explanation of the adding algorithm

In Fig. 4, there are three cases of adding a translation pair. We compute the score $\tau(e_k, t_l)$ by using Eq.1 where the value of α is 0.8 for these three cases in normalizing step. In Fig. 4(a) is a case of unseen pairs. We add the pair with its probability to the original directly. In Fig. 4(b), the parent node is the same as one in original tree but the child node is different. We attach the child node to the original parent with its probability. Lastly, in Fig. 4(c) there is the same pair in the original graph; we only update the score without adding a new child node.

3.3 Phrase Level Combining

In order to recover missing phrases, we combined phrase translation models of the forward and backward training. These phrase-based translation models are combined using the adding algorithm as mentioned in Section 3.1. Instead of a word, each node in a graph contains a phrase.

3.4 Lexical Weight Calculation

Lexical weight [2] was used to assign a validation score to a phrase translation pair with a given alignment. The lexical weight is computed by Eq.3.

$$p_w(\bar{e} \mid \bar{t}, a) = \prod_{i-1}^{n} \frac{1}{|\{ j \mid (i, j) \in a \}|} \sum_{\forall (i,j) \in a} w(e_i \mid t_j) \tag{4}$$

Where English phrase $\bar{e} = e_1 e_2 ... e_n$ and Thai phrase $\bar{t} = t_1 t_2 ... t_m$ and a is a given alignment. $w(e_i \mid t_j)$ is the word-based translation probability. Fig. 5 shows a phrase translation pair $\bar{e} = e_1 e_2 e_3$, $\bar{t} = t_1 t_2 t_3$ with a given alignment.

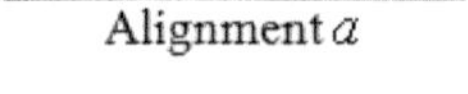

$$p_w(\bar{e} \mid \bar{t}, a) = p_w(e_1 e_2 e_3 \mid t_1 t_2 t_3 t_4, a)$$
$$= w(e_1 \mid t_1)$$
$$\times \frac{1}{2}(w(e_2 \mid t_2) + w(e_3 \mid t_2))$$
$$\times w(e_3 \mid t_4)$$

Fig. 5. Lexical weight calculation

Our method use Eq.4 to compute the lexical weight. But, instead of using the original word alignment model extracted by using the IBM Model-4, we use the combined word-based translation model in Section 3.1.

Fig. 6 shows the overview of our algorithm to construct phrase translation parameters of an English-to-Thai phrase pair.

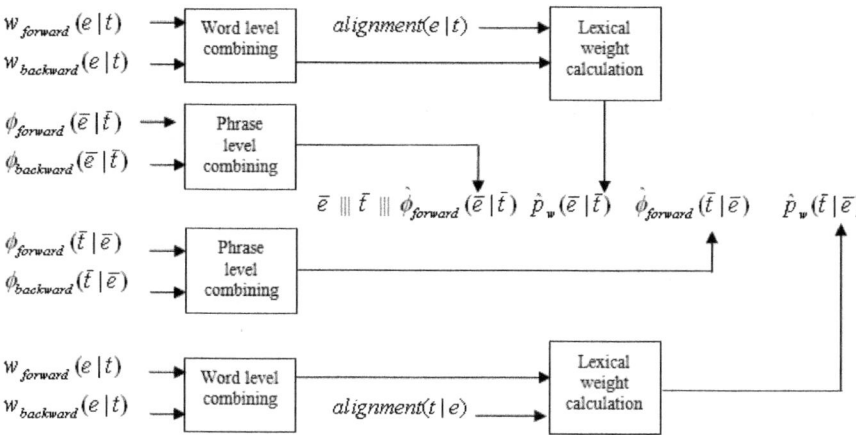

Fig. 6. The overview of our combining algorithm

4 Experiments

In the experiments, we use our English-Thai parallel corpus consisting of 200,000 sentence pairs. There are approximately 1.2 million words in the corpus. Thai sentences were lexically segmentedby our word segmentation toolkit called jWordSeg (http://www.suparsit.com/nlp-tools). 135 sentences whose lengths exceed 40 words are excluded fromtheexperiment The experiments use GIZA++ toolkit to align word pair SRILM [6] is used for constructing language model.Finally, Moses is used for decoding in order to produce translation results. The difficulties of detecting *missing phrases* during the training and test set separation encourage us to use a training set fora test set because we aim to investigate the number of *missing phrases*.

The Baseline Model
We alternatively trained all data in the corpus for English-to-Thai baseline model without dividing training set and test set in order to compare the performance of the baseline and improved training processes. To prepare data, we train English-to-Thai phrase translation table by using GIZA++ and   phrase extraction algorithm based described in Section 2.

The Combined Models
According to the algorithm, we design the weight combining value α in the word level combining and the phrase level combining processes. In our experiment, we manually adjust the combining weight from 0.1 to 0.9.

   In decoding step, we set weight parameters for each score equally (0.2 as default). The weight of language model is set to 0.5 as default. For the baseline and the combined models, we evaluate the system by calculating the BLEU score [7] and the NIST score [8]. These scores show how similarity of translation result between machine and human based on an n-gram approach.

5 Results and Discussion

Table 1 shows BLEU scores and NIST scores of the baseline model and combined models. It is obvious that the recovery of missing phrases shows better results. The combined model with 0.1 of α yields the best score. In average, about 1.3% of BLEU scores and 0.9 % of NIST scores are increased.

Table 1. BLEU scores and NIST scores of the baseline model

Model	α	BLEU	NIST
Baseline	-	0.5492	13.3945
Combined#1	0.1	0.5562	13.5214
Combined#2	0.2	0.5562	13.5202
Combined#3	0.3	0.5561	13.5187
Combined#4	0.4	0.5560	13.5174
Combined#5	0.5	0.5560	13.5173
Combined#6	0.6	0.5559	13.5152
Combined#7	0.7	0.5559	13.5152
Combined#8	0.8	0.5559	13.5133
Combined#9	0.9	0.5559	13.5128

There are 9,336 missing phrases in the baseline result. However, the number of missing phrases in combined models is reduced to 8,646 and 7.36% of *missing phrases* (690/9336) are recovered from baseline model. We extracted the result from Combine#1 model and compared it to Baseline model line-by-line. We realised that over 40,000 sentences were changed.

Fig. 7(a) obviously shows that the missing phrase *"summarize"* is recovered by our method. In Fig. 7(b), the baseline model cannot extract the phrase *"workload"*, while the Combine#1 model gives the translation result as *"ภาระ งาน นี้"*. The phrase pair *"workload-ภาระ งาน นี้"* was extracted from other sentences. Actually, the phrase *"ภาระ งาน นี้"* and *"ปริมาณ งาน นี้"* (in the desired target) shares the same meaning with *"workload"*.

In Fig. 7(c), the phrase *"has been vacant"* is successfully translated to *"มี ที่ ว่าง"* in baseline model. However, the Combined#1 model alternatively translates it to *"ได้ ว่าง"* which is exactly the desired translation results. Since the baseline model cannot extract *"has been vacant-ได้ ว่าง"*, the phrase *"has been vacant"* is also a *missing phrase*. Comparing to explicitly missing phrases, this type of missing phrases is different since it cannot be specified from the results directly because the decoder yields another alternative. This is separately named implicitly *missing phrase*. The recovery of hidden missing phrases gives significant improvements.

(a)	Sentence
Source	He *summarized* the whole proposal in three sentences .
Desired target	เขา *สรุป* ข้อเสนอ งาน ทั้งหมด ไว้ ใน สาม ประโยค
Baseline	เขา*summarized*ทั้ง ข้อเสนอ ใน สาม ประโยค นี้
Combine#1	เขา *สรุป* ข้อเสนอ งาน ทั้งหมด ไว้ ใน สาม ประโยค

(b)	Sentence
Source	The *workload* should be shared among the staff .
Desired target	*ปริมาณ งาน นี้* ควรจะ มี การแบ่ง กัน ทำ ใน หมู่ คณะ ทำงาน
Baseline	*พวก workload*ควร แบ่ง ใน หมู่ พนักงาน
Combine#1	*กระ งาน นี้* ควรจะ มี การแบ่ง กัน ใน หมู่ พนักงาน

(c)	Sentence
Source	This post *has been vacant* for three months .
Desired target	งาน ตำแหน่ง นี้ *ได้ ว่าง* มา สาม เดือน แล้ว
Baseline	งาน ตำแหน่ง นี้ *มี ที่ ว่าง* เป็นเวลา สาม เดือน แล้ว
Combine#1	งาน ตำแหน่ง นี้ *ได้ ว่าง* เป็นเวลา สาม เดือน แล้ว

Fig. 7. Examples of improved result compared to baseline

In summary, over 30,000 phrase pairs are recovered into the new phrase translation table from combined model.

6 Conclusions and Future Work

In this paper, we propose a method to recover missing phrase alignments from the *forward* phrase table and the *backward* phrase table. The phrase translation tables extracted from the source-to-target training and the target-to-source training are combined. A phrase translation table consists of several scores, for instance phrase translation probabilities and lexical weights. In order to calculate lexical weights, the forward and backward word translation models are merged. Next, the adding algorithm is used in both word and phrase level combining process with a specified combining weight.

We designed our experiments by adjusting the combining weight. We tested our combined model by using the training corpus in order to see how many missing phrases can be recovered. According to the results, 7.3% of unknown phrases were solved. 1.3% of BLEU score and 0.9% of NIST score are increased. Approximately 30,000 phrase pairs are increased in the combined phrase translation table.

Although our method is initially designed for combining the phrase translation model in a same corpus, we plan to extend this work to adjust combining weight from different corpora.

The ambiguity on word alignment leads to inconsistent phrase pair; therefore, in the future work, we will focus on word alignment model for sparse corpus. To investigate the effect on forward and backward model, we plan to experiment with different language pairs.

References

1. Brown, P.F., Stephen, A.D.P., Vincent, J.D.P., Robert, L.M.: The Mathematics of Statistical Machine Translation: Parameter Estimation. Computational Linguistics 19, 263–311 (1993)
2. Koehn, P., Franz, J.O., Daniel, M.: Statistical Phrase-Based Translation. In: Human Language Technology conference/North American chapter of the Association for Computational Linguistics (2003)
3. Hideo, O., Yamamoto, H., Sumita, E.: Introducing Translation Dictionary Into Phrase-based SMT. In: Proceedings of Machine Translation Summit, pp. 361–367 (2007)
4. Amittai, A.: Factored Language Model for Statistical Machine Translation. MRes Thesis. Edinburgh University (2006)
5. Koehn, P., Hoang, H., Birch, A., Callison-Burch, C., Federico, M., Bertoldi, N., Cowan, B., Shen, W., Moran, C., Zens, R., Dyer, C., Bojar, O., Constantin, A., Herbst, E.: Moses: Open Source Toolkit for Statistical Machine Translation. In: Annual Meeting of the Association for Computational Linguistics (ACL), demonstration session, Prague, Czech Republic (2007)
6. Stolcke, A.: SRILM an Extensible Language Modeling Toolkit. In: International Conference on Spoken Language Processing (2002)
7. Papineni, K., Roukos, S., Ward, T., Zhu, W.-J.: BLEU: a method for automatic evaluation of machine translation. Technical Report RC22176 (W0109-022), IBM Research Report (2001)
8. Doddington, G.: Automatic evaluation of machine translation quality using n-gram co-occurrence statistics. In: Proceedings of ARPA Workshop on Human Language Technology (2002)

Automatic Extraction of Thai-English Term Translations and Synonyms from Medical Web Using Iterative Candidate Generation with Association Measures

Kobkrit Viriyayudhakorn[1], Thanaruk Theeramunkong[1],
Cholwich Nattee[1], Thepchai Supnithi[2], and Manabu Okumura[3]

[1] Sirindhorn International Institute of Technology, Thammasat University
131 Moo 5, Tiwanont Rd., Bangkadi, Muang, Pathumthani, 12000, Thailand
Tel: +66 (0) 2501 3505-20, Fax: +66 (0) 2501 3524
[2] National Electronics and Computer Technology Center
112 Paholyothin Road, Klong 1,
Klongluang, Pathumthani, 12120, Thailand
[3] Precision and Intelligence Laboratory
Tokyo Institute of Technology
4259 Nagatsuta Midori Yokohama 226-8503, Japan
{kobkrit,cholwich,thanaruk}@siit.tu.ac.th,
thepchai.supnithi@nectec.or.th,oku@pi.titech.ac.jp

Abstract. Electronic technical documents available on the Internet are a powerful source for automatic extraction of term translations and synonyms. This paper presents an association-based approach to extract possible translations and synonyms by iterative candidate generation using a search engine. The plausible candidate pairs can be chosen by calculating their co-occurring statistics. In our experiment to extract Thai-English medical term pairs, four possible alternative associations; namely confidence, support, lift and conviction, are investigated and their performances are compared by ten-fold cross validation. The experimental results show that lift achieves the best performance with 73.1% f-measure with 67% precision and 84.2% recall on translation pair extraction, 68.7% f-measure with 71.5% precision and 67.7% recall on Thai synonym term extraction and 72.8% f-measure with 72.0% precision and 75.1% recall on English synonym term extraction. The precision of our approach in Thai-English translation, Thai synonym and English synonym extraction are 4 times, 3.5 times and 5.5 times higher than baseline precision respectively.

Keywords: Association rule mining, Thai-English medical term translation, Conviction, Lift, Iterative approach, Synonym extraction.

T. Theeramunkong et al. (Eds.): PAKDD Workshops 2009, LNAI 5669, pp. 141–155, 2010.
© Springer-Verlag Berlin Heidelberg 2010

1   Introduction

In recent years, there have been several attempts to extend text mining techniques to mine specific health-related knowledge, such as medical, pharmaceutical and biological [1,2,3] practices. Since health-related articles usually include a lot of technical terms, processing such terminologies becomes an important factor towards success of automating analysis of those articles. There are still a lot of challenges in developing Thai health-related terminology due to at least two reasons. First, currently there has been no standardization of health-related terminology in Thai languages. Second, it is a backbreaking task to add new terms or to modify information of terms in a conventional paper-based or online dictionary. Nowadays, since there are a lot of web pages providing information or knowledge related to health science, it is possible to use such pages as resources to construct health-related terminology. Normally, like texts in several non-English languages, Thai medical texts often include Thai technical terms followed by their corresponding English translations since English is widely recognized as a common language for interchanging technical information. Among several patterns of translation pairs, a common one is that the English translation of a Thai term is enclosed in a parenthesis or placed immediately after a term. For example, 'โรคกระเพาะ (peptic ulcer)' or 'โรคกระเพาะ peptic ulcer' denotes that the Thai word 'โรคกระเพาะ' has the term 'peptic ulcer' as its English translation.

This common regularity enables us to extract Thai-English translation pairs. However, there have been a few difficulties in extracting translation pairs from texts. First, a technical term may be translated into several different terms due to lacking of standardization. For instance, a term 'peptic ulcer' can be translated into four Thai translation terms, (1)'โรคกระเพาะอาหาร', (2)'โรคแผลในกระเพาะอาหาร', (3)'แผลเป็บติก' or (4)'แผลเพ็บติก' where the first two terms are direct translation, and the last two terms are transliteration. While an English term can be translated to more than one Thai terms, a Thai term is also able to be mapped to several English terms. Therefore, a mechanism to select the best translation pair is needed. Second, due to authors' writing styles, an English term after a Thai term may not be its translation. As one example, in Thai medical texts, sometimes an English term is used directly as a word in a context, without specifying its corresponding Thai term, such as 'สามารถทำให้เกิด dengue fever' ('can trigger dengue fever'). This irregularity causes difficulty in detecting translation pairs.

Intuitively it is possible to detect synonyms by linking two translation candidate pairs. To create a candidate for Thai-Thai synonym, we link a Thai-English translation pair with another English-Thai translation pair that has identical English term. In the same manner, linking a English-Thai translation pair and a Thai-English translation pair will enable us to obtain a candidate for English-English synonyms.

Towards the above objectives, this paper presents a method to use Web documents as resources for extracting translation and synonym pairs of English and Thai. This paper is organized as follows. Some previous approaches are described in Section 2. Section 3 presents the framework and techniques for constructing translation and synonym pairs. In Section 4, the experimental results are discussed. Finally, a conclusion and some further works are given in Section 5.

2 Related Works

This section gives a survey to research works related to extraction of term translations and synonyms. Some recent works have been conducted to extract term pairs between English and Chinese translations from Chinese texts on the Internet. In [2], Zhang and Vines proposed a method that generated English-Chinese and Chinese-English translation candidates from top-100 search results from a search engine and then find potential translation pairs by exploiting co-occurrence frequency and surface characteristics, such as term length and common substring.

As another work, Wang and his colleagues [4,5,6] showed promising results of exploiting the Web as a source to generate effective translation equivalents for many unknown terms, including proper nouns, technical terms and Web query terms and in assisting bilingual lexicon construction for a real digital library system. Their method applied the Chi-square test and context vector analysis to rank cohesion among terms in web documents to tackle with low-frequency problem.

English synonym extraction by using an unsupervised learning algorithm based on statistical data was proposed in [7] and was improved by combining with symbolic knowledge in [8]. For Japanese language, Okamoto[9] extracted a set of near-synonyms by using semantic features from the Thesaurus and then weighting them with their occurrence probabilities under a set of heuristics. Shimohata[10] extracted synonyms from documents whose contents are similar by looking contextual information of surrounding words. In Thai language, the number of works in mining translation pairs is still limited. As our best knowledge, there is still no work on Thai synonym extraction using web documents.

3 Iterative Approach for Candidate Generation and Association-Based Candidate Selection

In this section, an approach to extract translation pairs and synonyms using iterative candidate generation and association-based selection is described. The two main steps in our approach are (1) iterative candidate generation and (2) candidate selection.

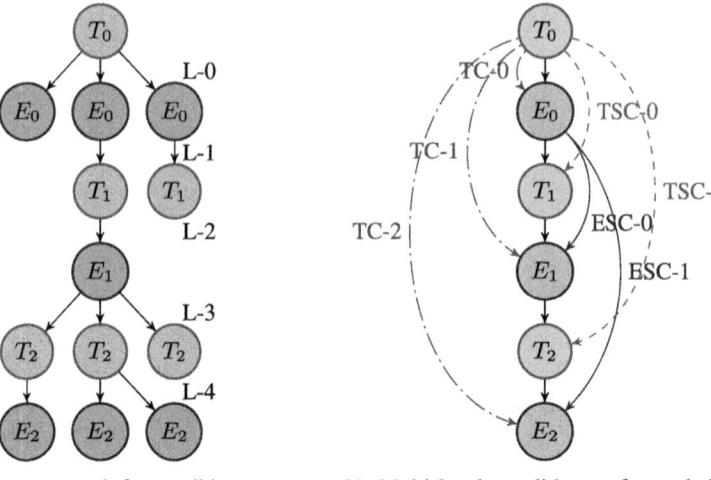

(a) Iterative approach for candidate generation

(b) Multi-level candidates of translation pairs and synonym pairs

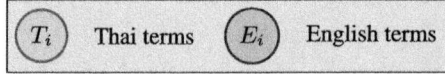

Fig. 1. Mapping candidate generation and multi-level candidates of translation pairs and synonym pairs (L-n: n^{th} iteration, TC-n: The nth-level mapping of Thai-English translation pairs, TSC-n: The nth-level mapping of Thai synonym pairs, ESC-n: The nth-level mapping of English synonym pairs)

In the first step, an initial set of Thai terms are fed to a search engine to get their potential English translations from Thai web documents. Next, these obtained English translations are submitted again to the search engine in order to retrieve their potential Thai translations from Thai web documents. This process is iteratively executed by inputting the obtained Thai translations into the search engine repeatedly to obtain English translations. In general, it is possible to specify the language of search result pages returned from the search engines, such as Google[4] and Yahoo[5]. With this feature, we have chosen Thai as the language of our target pages for all iterations. The search results are used as a corpus to extract term translations and synonyms.

In the second step, after a number of potential translations are obtained, an association-based measure is applied to select the most plausible candidates as translation and synonym pairs. Figure 1 illustrates the whole process where the details are shown in the following subsections.

---

[4] http://www.google.com
[5] http://www.yahoo.com

3.1   Iterative Candidate Generation

As the first step, our proposed approach starts from a set of initial Thai technical terms and then find their English translations. Each Thai term is submitted as a query to a search engine. Then, a set of the top-k result snippets are obtained from the search engine. The common writing patterns are applied to each snippet to extract English translations of the term. Then, each obtained English translation is re-submitted to the search engine and different Thai terms may be extracted from the results. The process is repeatedly conducted for each newly obtained English or Thai terms until no new term is obtained from the search results or the number of iterations is higher than a pre-defined threshold. The result of the iterative extraction for each term is represented as a directed graph as shown in Figure 1(a). Each node of the graph denotes a term. Each arc links a term with its translated term in another language.

The iterative approach allows us to construct a set of potential translation candidates since a Thai term can be translated into more than one English terms, or there are more than one Thai terms used to refer to an English term. For example, a Thai term 'โรคหูด' is a translation of both 'ringworm' and 'tinea' which are synonym. By linking two related term translation pairs, the proposed method can be also used to generate a pair of synonyms. In this case, a Thai term can be used as an intermediate to obtain a synonym of the given English term. In the same manner, we can link two Thai terms through a English term. This linking can also be done through more than one intermediates as shown in Figure 1(b) to obtain more than one synonyms.

Since, there is no explicit word boundary in the Thai writing system, it is possible to extract an incorrect portion (string) as a word in a running text. For example, we may get a phrase 'มีอาการคล้ายไข้เดงกี' ('has symptoms similar to dengue fever') that includes an useless part 'มีอาการคล้าย' ('has symptoms similar to') in front of a suitable word 'ไข้เดงกี' because, as occurred often in Thai natural writing style, a space may not be inserted between that part and the target term 'ไข้เดงกี'('dengue fever'). To filter out these incorrect pairs, we have proposed to use association-based candidate selection. The basic idea is that the incorrect pairs usually have low occurrence frequency. More details will be described in the next subsection.

3.2   Association-Based Candidate Selection

The association analysis aims to evaluate the relationship between two sets of objects (set A and set B) written as $A \rightarrow B$. The $A \rightarrow B$ indicates that B is likely to occur when A occurs. In our experiments, if A is set to be the source language term and B is set to be the target language term, we can use $A \rightarrow B$ to indicate how likely B will be taken place when A occured. In other words, how likely B is the translation of A when

B is a term that is enclosed by parenthesis or is placed immediately after A in common writing pattern. Normally, association is quantified by a set of well-known measures in association rule mining; namely support, confidence, lift and conviction. They have strong and weak points under different situations. Next, we explain how the association measures are applied to measure the association between terms in a translation pair.

$N(X)$ is the number of pages that include the word X and $N(*)$ is the total amount of existing pages. Unfortunately computing $N(*)$ is impossible since the total number of Web pages indexed by search engine are not precisely estimated. However $N(*)$ may be trivial when only ranking result is concerned.

- **Support** is an undirected measure that specifies the ratio that A and B occur with respect to the total occurrence.

$$Support(A \rightarrow B) = \frac{N(A \wedge B)}{N(*)} \tag{1}$$

- **Confidence** is a directed measure specifying the ratio that A and B occurs when A occurred.

$$Conf(A \rightarrow B) = \frac{N(A \wedge B)}{N(A)} \tag{2}$$

- **Lift** or **Interestingness** is an undirected measure that has an advantage over confidence by exploiting negative association. Lift measures the proportion of A and B occurring together compared to the expected occurrence when they are considered statistically independent.

$$Lift(A \rightarrow B) = \frac{N(*)N(A \wedge B)}{N(A)N(B)} \tag{3}$$

- **Conviction** is a directed measure representing the proportion of A occurrences without B, comparing to the expected occurrence when they are dependent. i.e. $N(A)$ and $N(\neg B)$.

$$Conv(A \rightarrow B) = \frac{N(A)N(\neg B)}{N(*)N(A \wedge \neg B)} \tag{4}$$

To calculate association measures, we submit queries to a search engine and use the number of page hits returned from the search engine as probability estimation. From the above association measures, the search results from the search engine for association measures are obtained by submitting both A and B for $N(A \wedge B)$, either A or B for $N(A)$ and $N(B)$, and both A and $-B$ for $N(A \wedge \neg B)$. B is leaded by the minus sign to specifies that B is an unwanted word.

However, computing conviction requires some assumptions since we cannot submit a query to obtain the web pages not containing B for finding $N(\neg B)$. Anyway, since the number of pages available on the Web is very large and B is a specific technical terms that are rarely found on the Web, we can assume that $\frac{N(\neg B)}{N(*)}$ is very closed to 1. Therefore, the approximated conviction can be computed as

$$Conv^*(A \rightarrow B) = \frac{N(A)}{N(A \wedge \neg B)} \tag{5}$$

As stated above, the association measures are of two types: directed and undirected measure. A directed measure evaluates the relationship of $A \rightarrow B$, (i.e. A causes occurrence of B), differently from the relationship $B \rightarrow A$. In contrast, the undirected association measures do not take into account the direction of occurrence. It measures both relationships in the same manner. Since we suppose that the relation between terms in the translation and synonym pairs are undirected relation, the directed association measure such as conviction and confidence need to be translated into an undirected representation. At this step, three functions; Minimum, Maximum and Mean are proposed to combine two directed measures to be an undirected one.

$$\lambda^{Min}(A \leftrightarrow B) = Min(\lambda(A \rightarrow B), \lambda(B \rightarrow A)) \tag{6}$$

$$\lambda^{Max}(A \leftrightarrow B) = Max(\lambda(A \rightarrow B), \lambda(B \rightarrow A)) \tag{7}$$

$$\lambda^{Mean}(A \leftrightarrow B) = Mean(\lambda(A \rightarrow B), \lambda(B \rightarrow A)) \tag{8}$$

where λ represents an directed association measure which is confidence or conviction, λ^{Min}, λ^{Max} and λ^{Mean} denotes an undirected association measure generated by Minimum function, Maximum function and Mean function respectively.

As an extension, association measures can be applied to find potential synonym pairs. However, a pair of synonyms are rarely occurred together on the same web page because a writer usually selects only one term to express each meaning in a sentence. Therefore, we cannot directly apply the search results to compute the measure for a pair of synonyms. However, with slight application, we can combine the association measures for term translation pairs in order to obtain a synonym pair.

In this paper, how likely two terms are a synonym pair is determined by considering the minimum association measures obtained from their translation intermediates. For example, while the arc labeled as 'TSC-1' in Figure 1(b) presents the example of a Thai-Thai synonym pair, T_0 and T_2, their association value is determined by the minimum value among the association values of all intermediate translation pairs, (T_0, E_0), (E_0, T_1), (T_1, E_1) and (E_1, T_2).

Iterative candidate generation repeatedly extracts the translated term from the product of the previous iteration. In any iteration when the association value is low, resulting in translated terms weaken the linkage in the next iteration. We infer that the association of synonym pairs is low when their intermediate linkage is weak. For this reason the minimum function is used for the association of the synonym pair $\mu_s(T_1, T_n)$ as defined below.

$$\mu_s(T_1, T_n) = \min_{i \in \{1...n-1\}} (\mu_t(T_i, T_{i+1})) \tag{9}$$

where T_1 and T_n represent a synonym pair. T_i and $T_i + 1$ denote an intermediate translation pair appearing between T_1 and T_n. μ_s represents an association measure used for evaluating each synonym pair. μ_t represents an association measure used for evaluating each translation pair.

4   Experiments

4.1   Experimental Settings

A number of experiments have been conducted to confirm the performance of the proposed approach. The prototype system is implemented based on the APIs provided by Yahoo Developer Center[6]. The APIs allows us to access and obtain XML results from the Yahoo search engine. An initial set of 510 Thai medical terms manually collected from various web pages are used as the initial set of terms for extracting translation. A set of English translation terms are generated from snippets returned from Yahoo APIs. The process is done repeatedly for five times as shown in Figure 1(a). We perform three experiments to solve three questions as follows.

In the first experiment, the performance of the three functions, namely Min, Max and Mean, used for combining two directed measures to be an undirected measure are compared to select the best one. The training set includes all translation pairs from iterative candidate generation. For each undirected confidence and undirected conviction, its score is ranked in the descending order and compared to another. Here, top-k word pairs are evaluated where k is varied from 1 to the number of all possible word pairs in order to evaluate which function is the best.

In the second experiment, we evaluate translation and synonym pairs obtained by four association measures i.e. support, confidence, lift and conviction. As directed association measures, confidence and conviction require a mechanism to combine two functions that are the result of the first experiment to be the undirected association measures. For each association measure, the top-k translation and synonym word pairs are displayed in the descending order when the result is output. The experiments are conducted to evaluate three levels of generated candidates for translation and two levels of

---

[6] http://developer.yahoo.com

generated candidates for synonym as shown in Figure 1(b). The results are compared with the results labeled by three human evaluators. When the evaluators give different labels on one word pair, majority voting is selected.

In the third experiment, the performance of four association measures is tested with an unseen test set. The ten-fold cross validation is applied. The translation and synonym candidate pairs are equally divided into ten parts. Nine parts stand as a training set and one remaining part is used as a test set with equal distribution among positive and negative examples. The test are conducted repeatedly for ten times. In each time, the test set is changed to another part that never tested before. For each association measure, the value of an association measure that yields the highest f-measure in the top-k training set will be used as the threshold in the test set. For each translation and synonym pair in test set, we have

$$C_{(x)} = \begin{cases} C^+ & \text{if } V_{(x)} \geq \delta, \\ C^- & \text{if } V_{(x)} < \delta. \end{cases}$$

where $C_{(x)}$ stands for the assigned class for a pair x in the test set, C^+ denotes the positive class, C^- denotes the negative class, $V_{(x)}$ represents the association value of the pair x in the test set, δ stands as a threshold in the test set. The performance of our proposed system is evaluated with false positive, precision, recall and f-measure. Here, let S be the set of generated word pairs, and C be the set of correct word pairs. We have

$$FalsePositive = \frac{|S - C|}{|S|} \tag{10}$$

$$Precision = \frac{|S \cap C|}{|S|} \tag{11}$$

$$Recall = \frac{|S \cap C|}{|C|} \tag{12}$$

$$F-measure = \frac{2 * Precision * Recall}{Precision + Recall} \tag{13}$$

4.2 Experimental Results

Table 1(a) shows the result of candidate generation with the number of input terms, the number of extracted terms, the ratio of extracted terms over input terms. It is found that candidate generation from Thai to English has branching factor of approximately 2 but approximately 0.75 in English to Thai. At the end of all iterations, we got 2,321 Thai words and 5,953 English words in total. Table 1(b) shows the precision of the extracted words is decreased in the later iteration.

Table 1. Basic characteristic of candidate generation and baseline precision of the translation and synonym pairs

(a) Candidate generation result

Iteration	Input terms	Extracted terms	Ratio(Extracted/Input)
level 0 ($T \rightarrow E$)	510	1129	2.214
level 1 ($E \rightarrow T$)	1129	991	0.878
level 2 ($T \rightarrow E$)	991	2024	2.042
level 3 ($E \rightarrow T$)	2024	1330	0.657
level 4 ($T \rightarrow E$)	1330	2800	2.105

(b) Baseline precision

Evaluation level	Baseline Precision(%)		
	Translation	Thai synonym	English synonym
Level 0	63.77	35.62	24.95
Level 1	11.96	6.39	0.42
Level 2	0.57	-	-
Total	16.43	18.87	12.40

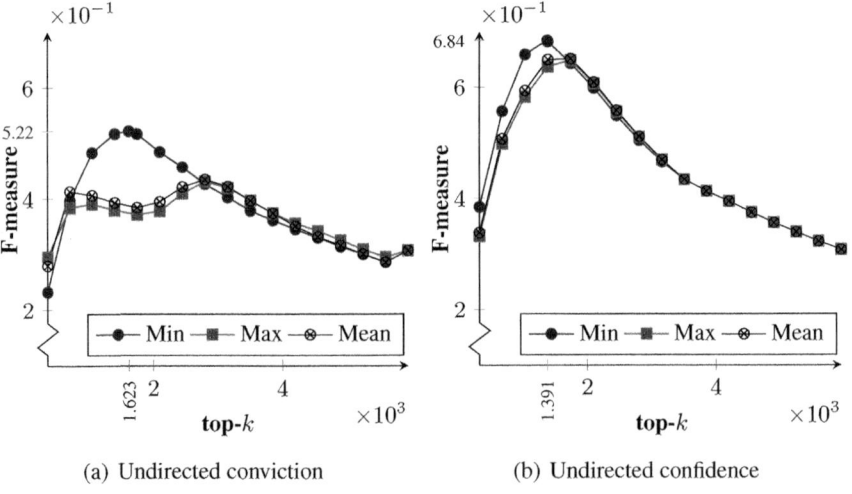

(a) Undirected conviction   (b) Undirected confidence

Fig. 2. F-measures of the top-k translation pairs in descending order (undirected confidence and undirected conviction). Three conditions considered are minimum, maximum and mean combining function.

In the first experiment, Figure 2 shows the f-measure of the top-k translation pairs of undirected confidence and undirected conviction that are combined by minimum, maximum and mean functions as shown in Section 3.2. The minimum function yielded

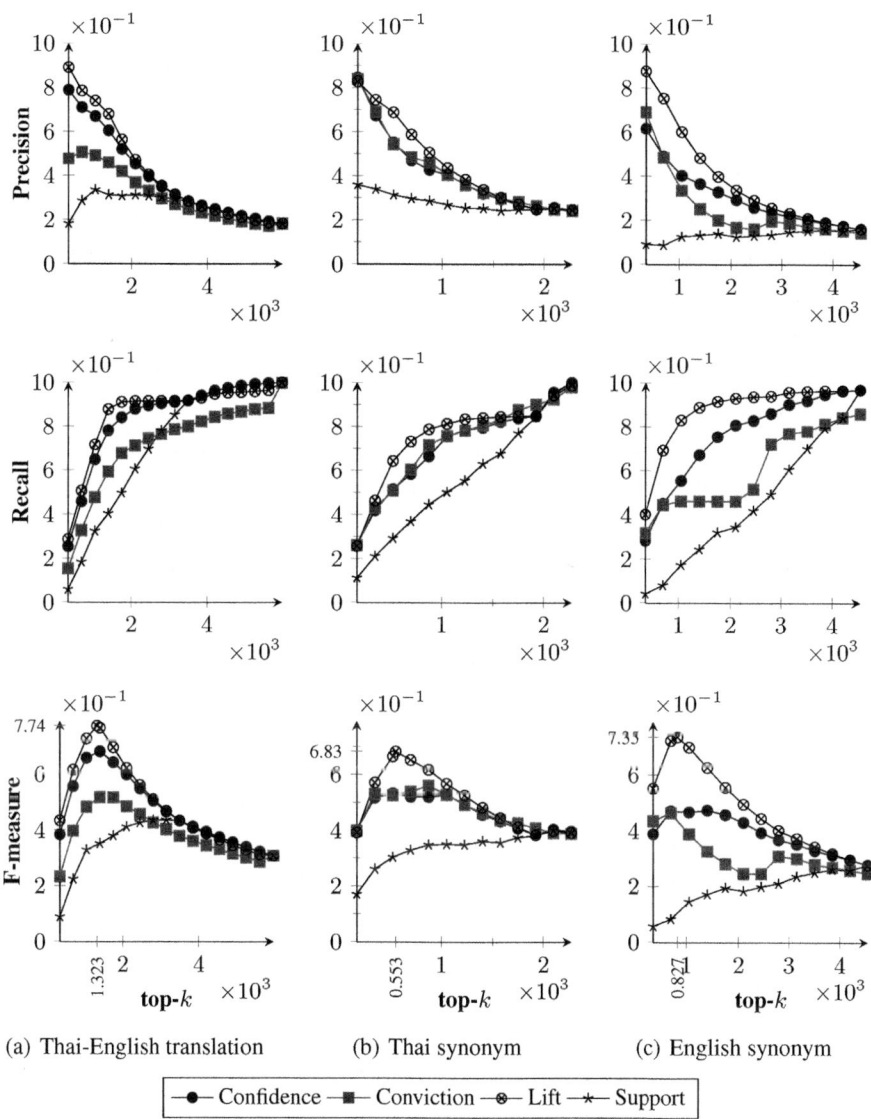

Fig. 3. Precision, Recall and F-measure for Thai-English Translation, Thai Synonym and English Synonym. Four association measures are Confidence, Conviction, Lift, Support.

the highest 68.44% f-measure in confidence and the highest 52.23% f-measure in conviction. Hence, we use the minimum function to generate undirected measures in the second and third experiment.

The second experiment investigates the precision, recall and f-measure score of top-k ranked by each association measures using all candidates as the training set and test set. Figure 3(a), 3(b) and 3(c) show that lift yields the best association measure with the highest 77.4% f-measure (with 70.3% precision and 85.9% recall) for Thai-English translation, the highest 68.3% f-measure (with 68.9% precision and 67.7% recall) for Thai synonym and the highest 73.4% f-measure (with 70.6% precision and 76.6% recall) in English synonym, respectively.

Figure 3(a) shows the precision, recall and f-measure score of top-k Thai-English translation pairs that rank by each association measures using all evaluation levels as training and test set. The lift×confidence achieved the best association measure with the highest 78.7% f-measure (with 70.8% precision and 83.5% recall) at top-1279. The precision of lift×confidence is approximately four times greater than baseline precision as shown in Table 1(b).

Figure 3(b) shows the precision, recall and f-measure score of top-k Thai synonym pairs that rank by each association measures in all evaluation levels. The highest 68.3% f-measure (with 68.9% precision and 67.7% recall) at top-553 obtained by lift. Its precision outperforms baseline precision as shown in Table 1(b) with three and a half times improvement.

For English synonym pairs, figure 3(c) shows the precision, recall and f-measure score of top-k English synonym pairs that rank by each association measures in all evaluation levels. Lift yielded the highest 73.5% f-measure at top-827 (with 70.6% precision and 76.6% recall) which is superior to baseline precision as shown in Table 1(b) with approximately five and a half times improvement.

As the result of the third experiment, Table 2 shows the average false positive rate, average precision, average recall and average f-measure of ten-fold cross validation (subscripted, with '10') when our proposed algorithm is applied. The last column of the table is the f-measure of our proposed algorithm using all candidates as the training test and test set (subscripted with 'all'). The difference between both f-measures is trivial. This means the stability of the proposed system.

Table 2(a) shows lift yields the highest 73.1% average f-measure (with 8.3% average false positive, 67.0% average precision and 84.2% average recall) for Thai-English translation. Its average precision is superior to baseline precision (16.43%) as shown in Table 1(b) with approximately 4 times improvement. Table 2(b) shows that lift yields the highest 68.7% average f-measure (with 9.9% average false positive, 71.5% average precision and 67.7% average recall) for Thai synonym. Its average precision outperforms baseline precision (18.87%) as shown in Table 1(b) with 3.5 times improvement. Table 2(c) shows that lift yields the highest 72.8% f-measure (with 6.0% false positive, the 72.0% precision and 75.1% recall) for English synonym. The average precision of lift is approximately 5.5 times greater than baseline precision (12.40%) as shown in Table 1(b).

Table 2. Experimental Result in Ten-fold Cross Validation

(a) Thai-English Translation

Model	False Positive$_{10}$	Precision$_{10}$	Recall$_{10}$	F-measure$_{10}$	F-measure$_{all}$
Lift	0.083	0.670	0.842	**0.731**	**0.774**
Confidence	0.112	0.599	0.777	0.654	0.684
Conviction	0.230	0.398	0.703	0.493	0.522
Support	0.438	0.296	0.831	0.423	0.442

(b) Thai synonym

Model	False Positive$_{10}$	Precision$_{10}$	Recall$_{10}$	F-measure$_{10}$	F-measure$_{all}$
Lift	0.099	0.715	0.677	**0.687**	**0.683**
Confidence	0.155	0.544	0.537	0.531	0.548
Conviction	0.461	0.396	0.712	0.464	0.561
Support	1.000	0.243	1.000	0.390	0.406

(c) English synonym

Model	False Positive$_{10}$	Precision$_{10}$	Recall$_{10}$	F-measure$_{10}$	F-measure$_{all}$
Lift	0.060	0.720	0.751	**0.728**	**0.735**
Confidence	0.095	0.531	0.493	0.487	0.498
Conviction	0.089	0.515	0.446	0.453	0.477
Support	0.924	0.164	0.966	0.280	0.281

Concludingly, the experimental results evidenced that our system with lift gained the highest f-measure (73.1%) for mining Thai-English translation pairs, compared to the extraction of English synonym pairs (72.8%) and Thai synonym pairs (68.7%). This result is quite intuitive since naturally it is necessary to extract at least two translation pairs to obtain a synonym pair. Moreover, comparing to mining of English synonyms, extracting Thai synonyms is a harder task since Thai language has no explicit word boundary.

5   Conclusion

This paper presented a method to use Web documents as resources for extracting translation and synonym pairs between English and Thai medical terms. Iteratively inputting keywords on a search engine, a set of translation candidate pairs are generated. The potential scores of translation word pairs are calculated using four alternative measures, support, confidence, lift and conviction, commonly used in association rule mining.

By experiments using 510 Thai words, we found out that our approach using lift as association measure achieves the highest average f-measure of ten-fold cross validation that is 73.1% (with 67% precision and 84.2% recall) for Thai-English translation, 68.7% (with 71.5% precision and 67.7% recall) for Thai synonym and 72.8% (with 72% precision and 75.1% recall) for English synonym. The precision of our approach in Thai-English translation, Thai synonym and English synonym are 4 times, 3.5 times and 5.5 times greater than baseline precision respectively. Lift is the best association measure for extracting both translation and synonym. We also found that the minimum function is the best function for combining two directed measure to be an undirected measure.

As our future work, we plan to improve our approach to using combination of association measures with larger data sets and different specific domains.

Acknowledgment

This research is financially supported by the Thailand Research Fund (TRF), under Grant No. BRG50800013 and NCRT Grant 2009 and Thailand Advanced Institute of Science and Technology - Tokyo Institute of Technology (TAIST-Tokyo Tech), National Science and Technology Development Agency (NSTDA), Tokyo Institute of Technology (Tokyo Tech) and Sirindhorn International Institute of Technology (SIIT), Thammasat University (TU).

References

1. Bodenreider., O.: Lexical, terminological, and ontological resources for biological text mining. In: Ananiadou, S., McNaught, J. (eds.) Text Mining for Biology and Biomedicine, ch. 3, pp. 43–66. Artech House (2006)
2. Zhang, Y., Vines, P.: Using the web for automated translation extraction in cross-language information retrieval. In: Proceedings of the 27th Annual International ACM SIGIR Conference (SIGIR 2004), Sheffield, South Yorkshire, UK, July 2004, pp. 162–169 (2004)
3. Viriyayudhakorn, K., Theeramunkong, T., Nattee, C.: Mining translation pairs for thai-english medical terms. In: Proceedings of the 3rd International Conference on Knowledge, Information and Creativity Support Systems (KICSS 2008), December 2008, pp. 104–111. Hanoi National University of Education (HNUE), Hanoi (2008)
4. Wang, J.-H., Teng, J.-W., Cheng, P.-J., Lu, W.-H., Chien, L.-F.: Translating unknown cross-lingual queries in digital libraries using a web-based approach. In: Proceedings of the 2004 Joint ACM/IEEE Conference on Digital Libraries (JCDL 2004), Tucson, Arizona, USA, June 2004, pp. 108–116 (2004)
5. Lu, W.-H., Lin, S.-J., Chan, Y.-C., Chen, K.-H.: Semi-automatic construction of the chinese-english MeSH using web-based term translation method. In: Proceedings of American Medical Informatics Association 2005 Symposium, pp. 475–479 (2005)
6. Wang, J.-H., Teng, J.-W., Lu, W.-H., Chien, L.-F.: Exploiting the web as the multilingual corpus for unknown query translation. J. Am. Soc. Inf. Sci. Technol. 57(5), 660–670 (2006)

7. Turney, P.D.: Mining the web for synonyms: Pmi-ir versus lsa on toefl. In: Flach, P.A., De Raedt, L. (eds.) ECML 2001. LNCS (LNAI), vol. 2167, pp. 491–502. Springer, Heidelberg (2001)
8. Inkpen, D.: A statistical model for near-synonym choice. ACM Trans. Speech Lang. Process. 4(1), 2 (2007)
9. Okamoto, H., Sato, K., Saito, H.: Preferential presentation of japanese near-synonyms using definition statements. In: Proceedings of the second international workshop on Paraphrasing, vol. 16, pp. 17–24 (2003)
10. Shimohata, M., Sumita, E.: Acquiring synonyms from monolingual comparable texts. In: Dale, R., Wong, K.-F., Su, J., Kwong, O.Y. (eds.) IJCNLP 2005. LNCS (LNAI), vol. 3651, p. 233. Springer, Heidelberg (2005)

Accurate Subsequence Matching on Data Stream under Time Warping Distance

Vit Niennattrakul, Dechawut Wanichsan, and Chotirat Ann Ratanamahatana

Department of Computer Engineering, Chulalongkorn University
Phayathai Rd., Pathumwan, Bangkok 10330 Thailand
{g49vnn,g49dwn,ann}@cp.eng.chula.ac.th

Abstract. Dynamic Time Warping (DTW) distance has been proven to work exceptionally well, but with higher time and space complexities. Particularly for time series data, subsequence matching under DTW distance poses a much challenging problem to work on streaming data. Recent work, SPRING, has introduced a solution to this problem with only linear time and space which makes subsequence matching on data stream become more and more practical. However, we will demonstrate that it may still give inaccurate results, and then propose a novel Accurate Subsequence Matching (ASM) algorithm that eliminates this discrepancy by using a global constraint and a scaling factor. We further demonstrate utilities of our work on a comprehensive set of experiments that guarantees an improvement in accuracy while maintaining the same time and space complexities.

Keywords: Subsequence Matching, Data Stream, Dynamic Time Warping Distance.

1 Introduction

Dynamic Time Warping (DTW) distance measure, the distance that is widely used in various time series mining tasks, especially in classification [1], is largely established as one of the most accurate methods in finding similarity between two time series data. Past research work [2] has confirmed that the DTW distance measure dominantly outperforms the traditional Euclidean distance metric in terms of accuracy since the DTW distance measure exploits dynamic programming that allows more flexibility in sequence alignments. However, it relatively requires much higher time and space complexities.

Especially in working on data streams, DTW at first seems unattainable. In subsequence matching problems, we try to find most similar subsequences to the query on a much longer candidate sequence, i.e., a data stream. A candidate sequence can either be a fixed-length sequence or an infinite streaming sequence, producing a much more challenging problem. A brute-force method for subsequence matching simply extracts every subsequence from the candidate sequence, and then these sequences are all compared with the query sequence in a similar fashion to the whole sequence matching approach. This clearly is impractical for large or streaming data.

T. Theeramunkong et al. (Eds.): PAKDD Workshops 2009, LNAI 5669, pp. 156–167, 2010.

SPRING [3], a recently proposed subsequence matching algorithm on a data stream under DTW distance, has been introduced to provide solutions for the best-matched query and range query. SPRING algorithm obtains the same retrieval results as the brute-force subsequence matching approach with an impressively low complexity. Unfortunately, we have discovered that SPRING may give undesired results. To illustrate our point, Figure 1 shows an example of this discrepancy in the SPRING algorithm that fails to accurately retrieve the best-matched subsequence.

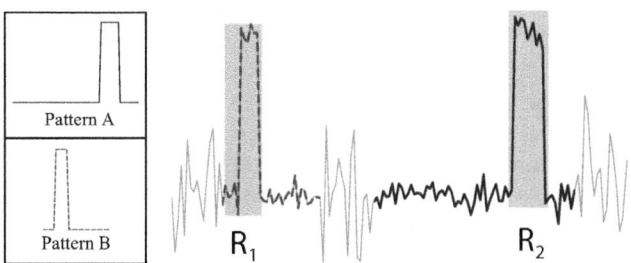

Fig. 1. Illustration of failure in retrieving accureate subsequences of SPRING

Given two pattern sequences of interest, A and B, which have similar shape, but with different lengths, onsets, and offset positions, we would like to find the best-matched subsequences to each pattern on a streaming sequence $\hat{S}$. If we use pattern A as a query, R_1 (in left highlighted box) will be incorrectly retrieved as the best match, and both R_1 and R_2 (in right highlighted box) will be incorrectly detected in the range query. In fact, we expect that if we use pattern A as a query, only a blue solid sequence will be an answer for both range query and the best-matched query. On the other hand, if we use pattern B as a query, though R_1 will be incorrectly retrieved as the best match due to its shorter length, both R_1 and R_2 will again be incorrectly detected for the range query. The correct result is expected to be a red dashed-dot sequence. Although this example is a bit contrived, it clearly demonstrates undesired results from SPRING algorithm which is critical to achieving an accurate subsequence matching algorithm.

In this work, we introduce ASM – Accurate Subsequence Matching – which is fast and accurate. We extend an idea of linear-time subsequence matching from SPRING, and we generalize the subsequence matching to support a global constraint and uniform scaling, where SPRING algorithm is ASM's special case. Our algorithm is comprehensively examined over 10 datasets to demonstrate effectiveness of our proposed work comparing with the best existing method, SPRING.

The remainder of this paper is organized as follows. In Section 2, we state subsequence matching problems and provide essential background knowledge. We describe our proposed work, ASM, in Section 3, and then report experimental results and give discussion in Section 4, before concluding our work in Section 5.

2   Problem Definition and Background

In this section, we review related work, define problems for subsequence matching, and provide background knowledge of Dynamic Time Warping distance measure, global constraints, and SPRING algorithm.

2.1   Related Work

With proven superiority of DTW distance measure, many subsequent works have been proposed to speed up its calculations by exploiting various indexing techniques through the use of lower bounds [4,5,6,7,8]. However, all of these techniques are designed for non-streaming data. Therefore, the advent of research on data stream has triggered a great number of works [9,10]. But, only until recently, SPRING algorithm [3] has been introduced to solve the subsequence matching problem on streaming data under DTW distance.

After the introduction of SPRING, many extensions and its applications [11,12,13] have been proposed, including Fast Subsequence Matching (FSM) [13] and Embedded Subsequence Matching (EBSM) [11]. More specifically, FSM extends SPRING to further reduce unnecessary distance calculations, and EBSM computes an approximate distance for subsequence matching by modifying query data and the data stream in vectors before the actual DTW calculations. However, all these works mainly focus on speed of the calculation, but not the retrieval accuracy. Therefore, this work attempts to improve retrieval accuracy without affecting the time and space complexities of the algorithm. We will first start by familiarizing the readers with formal problem definitions and essential background.

2.2   Problem Definition

In this section, we formalize and define two fundamental types of query – non-overlapping range query and non-overlapping top-k query – that are essential for the subsequence matching problem and the rest of this work. Let Q be a fixed-length query sequence with length n, $S[t_s : t_e]$ be a subsequence of data stream S from time t_s to t_e, R be a global constraint, and $[n_{min}, n_{max}]$ be the sequence length ranging from a scaling range of $[f_{min} : f_{max}]$. This scaling range is a user-defined parameter that indicates possible lengths of a candidate subsequence, where $n_{min} = f_{min} \times n$, $n_{max} = f_{max} \times n$, and n is the query sequence's length. When no global constraint and scaling are applied, non-overlapping range query is equivalent to SPRING algorithm's, and non-overlapping top-k query is equivalent to the best-matched query in SPRING when $k = 1$. In this work, we define $D_R(X, Y)$ as the DTW distance with a global constraint R of data sequences X and Y. In typical subsequence matching problem, a subsequence from non-overlapping range query is reported when a local minimum-distance subsequence is found, and subsequences from non-overlapping top-k query are reported when a set of top-k subsequences is changed.

Definition 1 (Non-Overlapping Range Query).   Non-overlapping range query returns a set $\Omega_{NORange}$ of non-overlapping subsequences $S[t_s : t_e]$ whose distance to a query sequence Q is less than a specific threshold ϵ under DTW with a global constraint R, and the length of a subsequence is between n_{min} and n_{max}.

Definition 2 (Non-Overlapping Top-k Query).   Non-overlapping top-k query returns a set Ω_{NOTopK} of first k non-overlapping subsequences $S[t_s : t_e]$ with smallest distances resulted from non-overlapping range query with constrained DTW measure. The lengths of subsequences are also between n_{min} and n_{max}.

2.3   Dynamic Time Warping Distance Measure

Dynamic Time Warping (DTW) distance measure [14,15] is a well-known shape-based similarity measurement. It uses a dynamic programming technique to find an optimal warping path between two time series sequences. Suppose we have two time series, a sequence $X = \langle x_1, x_2, \ldots, x_i, \ldots x_n \rangle$ and a sequence $Y = \langle y_1, y_2, \ldots, y_j, \ldots y_m \rangle$. The distance is calculated by following equations.

$$D(X_{1\ldots n}, Y_{1\ldots m}) = d(x_n, y_m) + \min \begin{cases} D(X_{1\ldots n-1}, Y_{j\ldots m-1}) \\ D(X_{1\ldots n}, Y_{1\ldots m-1}) \\ D(X_{1\ldots n-1}, Y_{1\ldots m}) \end{cases} \tag{1}$$

where $D(\emptyset, \emptyset) = 0$, $D(X_{i\ldots n}, \emptyset) = D(\emptyset, Y_{j\ldots m}) = \infty$, and $\emptyset$ is an empty sequence. Any distance metrics can be used for $d(x_i, y_j)$, including L_1-norm, $d(x_i, y_j) - |x_i - y_j|$, and L_2 norm, $d(x_i, y_j) = (x_i - y_j)^2$. For simplicity, we use L_1-norm to describe our proposed method, but L_2-norm is used in experimental evaluation to achieve better accuracy.

However, in reality, DTW measure may not give the best alignment that fits our need as it tries its best to find a minimum distance, it may generate an unwanted path. Without a global constraint, DTW measure will find its optimal mapping between the two time series data. We can resolve this problem by simply limiting the permissible warping paths using a global constraint.

2.4   Global Constraints

Although unconstrained DTW distance measure gives as optimal distance between two time series data, an unwanted warping path may be generated. The global constraint [1,16,17] efficiently limits the optimal path to give a more suitable alignment. Recently, Ratanamahatana-Keogh band (R-K band), a general model of global constraints, has been proposed, as shown in Figure 2. It can be specified by a one-dimensional array R, i.e., $R = \langle r_1, r_2, \ldots, r_i, \ldots, r_n \rangle$, where n is the length of time series, and r_i is the height above the diagonal in y direction and the width to the right of the diagonal in x direction. Each r_i value is arbitrary; therefore, R-K band is also an arbitrary-shaped global constraint. Note that when $r_i = 0$, where $1 \leq i \leq n$, this R-K band represents the classic Euclidean distance, and when $r_i = n$, $1 \leq i \leq n$, this R-K band represents the original DTW distance without constraint.

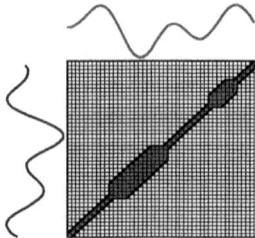

Fig. 2. An arbitrary-shaped global constraint, R-K band

2.5   SPRING Algorithm

SPRING algorithm [3], the first-proposed subsequence matching on data stream under DTW measure, can calculate optimal distance among subsequences in the data stream requiring $O(n)$ in both time and space complexities, where n is the length of a query sequence. SPRING is implemented based on two main ideas of Star-padding technique and STWM (Subsequence Time Warping Matrix). Star-padding is used to separate the overlapped subsequences, and STWM is a data structure that stores a minimum distance $d_{t,i}$ and a starting position $sp_{t,i}$. Suppose we have streaming sequence $S = \langle s_1, s_2, \ldots, s_t, \ldots \rangle$ and a query sequence $Q = \langle q_1, q_2, \ldots, q_i, \ldots, q_n \rangle$. At each time slice, new elements are calculated by following Equation 2 and Equation 3.

$$d_{t,i} = \|s_t - q_i\| + d_{best} \tag{2}$$

$$sp_{t,i} = \begin{cases} sp_{t-1,i-1} & \text{if } d_{best} = d_{t-1,i-1} \\ sp_{t,i-1} & \text{if } d_{best} = d_{t,i-1} \\ sp_{t-1,i} & \text{if } d_{best} = d_{t-1,i} \end{cases} \tag{3}$$

where $d_{t,0} = 0$, $d_{0,i} = \infty$, and $d_{best} = \min\{d_{t-1,i-1}, d_{t,i-1}, d_{t-1,i}\}$.

For more detail, a complete description of SPRING algorithm can be found in [3].

3   Accurate Subsequence Matching Algorithm

In this section, ASM (Accurate Subsequence Matching) algorithm and two important ideas, i.e., Scaled-Array (S-A) band and Modified Subsequence Matrix (MSM), are proposed. We describe S-A band and MSM in Sections 3.1 and 3.2, respectively. In Section 3.3, we introduce our novel subsequence matching algorithm, ASM, which supports two important features, i.e., a global constraint and scaling range.

3.1   Scaled-Array Band

The current global constraint representations are applicable only for a squared calculation matrix. Thus, we propose a Scaled-Array (S-A) band which can represent a global constraint as a single vector array A. More specifically, suppose we have an n-by-m path matrix, S-A band is defined as $A = \langle a_1, a_2, \ldots, a_i, \ldots, a_n \rangle = \langle (\alpha_1, \beta_1), (\alpha_2, \beta_2), \ldots, (\alpha_i, \beta_i), \ldots, (\alpha_n, \beta_n) \rangle$, where $1 \leq \alpha_i \leq \beta_i \leq m$ for all $1 \leq i \leq n$. Each element a_i in A collects a tuple of vector, i.e., a valid starting position α_i and a valid ending position β_i. A valid starting position is the smallest j that a cell (i, j) is a valid position within a global constraint, and a valid ending position is the largest j that a cell (i, j) is a valid position within a global constraint, where $1 \leq j \leq m$.

Since R-K band cannot be used as a global constraint for comparing time series data with different lengths, S-A band is proposed to represent global constraint. Suppose we have a query sequence with length n and a candidate sequence with length is a scaling range from f_{min} to f_{max}. Valid starting and ending positions (α_i, β_i) are defined in Equations 4 and 5.

$$\alpha_i = \arg \min \{k | k + r_k \geq i\} \times f_{min} \tag{4}$$

$$\beta_i = \begin{cases} (i + r_i) \times f_{max} & ; \text{if } i + r_i \leq n \\ (n - (i + r_i)) \times f_{max} & ; \text{otherwise} \end{cases} \tag{5}$$

where $0 < f_{min} \leq f_{max}$, $r_i \leq n$, $1 \leq i \leq n$, and $1 \leq \alpha_i \leq \beta_i \leq m$.

Note that a full global constraint and Euclidean distance are also defined when $\alpha_i = 1, \beta_i = \infty, 1 \leq i \leq n$, and $\alpha_i = i, \beta_i = i, 1 \leq i \leq n$, respectively.

3.2 Modified Subsequence Matrix

Modified Subsequence Matrix (MSM) is a data structure, derived from Subsequence Time Warping Matrix (STWM) in SPRING, where each element consists of four values, i.e., distance $d_{t,i}$, starting position $sp_{t,i}$, and positions x and y $(x_{t,i}, y_{t,i})$ on a global constraint S-A band. Each element (t, i) containing $d_{t,i}$, $sp_{t,i}$, $x_{t,i}$, and $y_{t,i}$ means that at this point, we have a valid subsequence from $sp_{t,i}$ that give the optimal distance $d_{t,i}$ at its coordinate $(x_{t,i}, y_{t,i})$ on a global constraint. A "valid" subsequence is defined as the subsequence that all coordinates $(x_{t,i}, y_{t,i})$ in its warping path from $sp_{t,i}$ to t are valid within the S-A band. Updating algorithm for an element in MSM is described in the next section.

3.3 Accurate Subsequence Matching

The basic idea behind ASM is the validation before an update of $d_{t,i}$ from $d_{t-1,i-1}$, $d_{t-1,i}$, or $d_{t,i-1}$ in MSM. We check the validity of global constraint position $(x_{t,i}, y_{t,i})$ from position $(x_{t-1,i-1} + 1, y_{t-1,i-1} + 1)$, $(x_{t-1,i} + 1, y_{t-1,i})$, and $(x_{t,i-1}, y_{t,i-1} + 1)$. If some positions $(x_{t-1,i-1}, y_{t-1,i-1})$, $(x_{t-1,i}, y_{t-1,i})$, or $(x_{t,i-1}, y_{t,i-1})$ make $(x_{t,i}, y_{t,i})$ invalid on the S-A band, these positions will not be selected in the calculation for d_{best}. Let $Q = \langle q_1, q_2, \ldots, q_i, \ldots, q_n \rangle$ be a query sequence, $S = \langle s_1, s_2, \ldots, s_t, \ldots \rangle$ be a streaming sequence, and $A = \langle (\alpha_1, \beta_1), (\alpha_2, \beta_2), \ldots, (\alpha_i, \beta_i), \ldots, (\alpha_n, \beta_n) \rangle$ be an S-A band with scaling ranges from f_{min} to f_{max}. We define function $v(x, y)$ to validate position (x, y) in the S-A band. It returns value 1 if (x, y) lies within the global constraint or $\alpha_y \leq x \leq \beta_y$; otherwise, it returns positive infinity. Four values at time t are updated according to the following equations.

$$d_{t,i} = \|q_i - s_t\| + d_{best} \tag{6}$$

$$sp_{t,i} = \begin{cases} sp_{t-1,i-1} & \text{if } d_{best} = d_{t-1,i-1} \\ sp_{t,i-1} & \text{if } d_{best} = d_{t,i-1} \\ sp_{t-1,i} & \text{if } d_{best} = d_{t-1,i} \end{cases} \tag{7}$$

$$x_{t,i} = \begin{cases} x_{t-1,i-1} + 1 & \text{if } d_{best} = d_{t-1,i-1} \\ x_{t,i-1} & \text{if } d_{best} = d_{t,i-1} \\ x_{t-1,i} + 1 & \text{if } d_{best} = d_{t-1,i} \end{cases} \tag{8}$$

$$y_{t,i} = \begin{cases} y_{t-1,i-1} + 1 & \text{if } d_{best} = d_{t-1,i-1} \\ y_{t,i-1} + 1 & \text{if } d_{best} = d_{t,i-1} \\ y_{t-1,i} & \text{if } d_{best} = d_{t-1,i} \end{cases} \tag{9}$$

where $d_{t,0} = 0$, $d_{0,i} = \infty$, and $d_{best} = \min \begin{cases} d_{t-1,i-1} \times v(x_{t-1,i-1} + 1, y_{t-1,i-1} + 1) \\ d_{t-1,i} \times v(x_{t-1,i} + 1, y_{t-1,i}) \\ d_{t,i-1} \times v(x_{t,i-1}, y_{t,i-1} + 1) \end{cases}$.

As we can see, values of each element at time t depends only on the previous element values at time $t-1$. Therefore, in practice, only two arrays are required. We denote d_i, sp_i, x_i, and y_i as distance, starting point, x position, and y position at current time t and at the query sequence position i, and d'_i, sp'_i, x'_i, and y'_i as distance, starting point, x position, and y position at previous time $t-1$.

For non-overlapping range query, a subsequence $S[t_s : t_e]$ is considered a result when minimum DTW distance with S-A band between $S[t_s : t_e]$ and Q is less than a threshold ϵ, and the subsequence has a qualified length. In addition, ASM will report this subsequence when $\forall i, d_i \geq d_{min} \vee sp_i > t_e$, as shown in Table 1. It is important to note that SPRING algorithm is a special case of our ASM when $(\alpha_i, \beta_i) = (1, \infty)$ for all i in the S-A band and a scaling length is $[1, \infty]$, so the same results are returned when non-overlapping range query is issued.

Non-overlapping top-k query is used to monitor a set of k subsequences which has minimum distance among overlapped subsequences. Generally, non-overlapping top-k query is implemented on non-overlapping range query whose initial threshold ϵ be positive infinity. When optimal range subsequence is found, we push this subsequence into a distance-priority queue. If size of the queue exceeds k, we pop the maximum-distance subsequence, and reset threshold ϵ to be a maximum distance of the queue. ASM algorithm for top-k query is shown in Table 2.

We would like to emphasize that ASM always achieves higher accuracy (as will be shown in our experimental section) while maintaining the same time and space complexities as SPRING's, i.e., $O(n)$ for both space and time complexities at each time slice, where n is the length of a query sequence. Since ASM keeps a single matrix and a single array which are both length n, and updates $O(n)$ numbers every time slice, ASM requires only $O(n)$ both in space and time.

Table 1. ASM algorithm for optimal range query

ALGORITHM ASMOptimalRange

---

1 Input: new streaming data point s_t
2 Output: optimal subsequence $S[t_s : t_e]$, if any
3 Let n be the length of a query sequence
4 $n_{min} = n \times f_{min}$, $n_{max} = n \times f_{max}$
5 for $i = 1$ to n do
6   COMPUTE d_i, sp_i, x_i, and y_i
7 endfor
8 if $d_{min} \leq \epsilon$
9   if $\forall i, d_i \geq d_{min} \lor sp_i > t_e$ then
10   REPORT(d_{min}, t_s, t_e)
11         $d_{min} = \infty$
12   for $i = 1$ to m do
13   if $sp_i \leq t_e$ then
14                 $d_i = \infty$
15   endif
16 endif
17 if $[d_m \leq \epsilon] \land [d_m < d_{min}] \land [n_{min} \leq x_m \leq n_{max}]$ then
19     $d_{min} = d_m$; $t_s = s_m$; $t_e = t$
20 endif
21 Substitute d'_i for d_i, sp'_i for sp_i, x'_i for x_i, y'_i for y;

Table 2. ASM algorithm for optimal top-k query

ALGORITHM ASMOptimalTopK

---

1 Input: new streaming data point s_t
2 Output: updated set P of top k
3 $S[t_s : t_e]$= ASMOPTIMALRANGE(s_t, ϵ)
4 if $(S[t_s : t_e] \neq NULL)$
5     $P.push(S[t_s : t_e])$
6   if $(size(P) > k)$
7         $P.pop()$
8         $e = P.peek().distance$
9   endif
10   REPORT(P)
11 endif

---

4 Experimental Evaluation

To evaluate the performance of our proposed method, we measure an accuracy of our algorithm comparing with the best existing algorithm, SPRING, using two evaluation metrics, i.e., Accuracy-on-Retrieval (AoR) and Accuracy-on-Detection (AoD).

4.1   Evaluation Metrics

We use two metrics, Accuracy-on-Retrieval (AoR) defined in Definition 3 and Accuracy-on-Detection (AoD) defined in Definition 4, to measure the quality of retrieval results on streaming sequences. Suppose we have a streaming sequence S, a set of expected pattern sequences E, and a set of retrieved sequences R. We first define an overlapping subsequence. Let $S[t_s : t_e]$ be the subsequence starting at t_s and ending at t_e. Overlapping subsequence $O_{X,Y}$ and overlap percentage $P_{X,Y}$, where $X = S[a : b]$ and $Y = S[c : d]$, are defined as follows.

$$O_{X,Y} = S[\max\{a,c\} : \min\{b,d\}] \tag{10}$$

$$P_{X,Y} = \frac{|O_{X,Y}|}{\max\{b,d\} - \min\{a,c\} + 1} \tag{11}$$

For instance, if we have subsequence $X = S[2 : 5]$ and $Y = S[3 : 7]$, $O_{X,Y} = S[3 : 5]$ and $P_{X,Y} = \frac{|S[3:5]|}{\max\{5,7\}-\min\{2,3\}+1} = \frac{3}{6} = 0.5$.

Definition 3 Accuracy-on-Retrieval. This evaluation measures how well an algorithm has *found* a set of expected subsequences while definition of *found* depends on overlapped percentage p. An extremely optimistic case is when p is 0, i.e., even subsequences are only one single data point overlapped, subsequence is marked as *found*. AoR is defined in Equation 12. Note that the higher the AoR, the better the result.

$$AoR = \frac{|\{O_{X,Y}|P_{X,Y} > p, X \in R, Y \in E\}|}{|E|} \tag{12}$$

Definition 4 Accuracy-on-Detection. This evaluation measures that once expected subsequences are *found*, as described in Definition 3, with overlapped percentage p, how well an algorithm can recognize these expected subsequences; in the other words, AoD is an average overlapping percentage $(P_{X,Y})$ of found sequences. AoD is defined in Equation 13. Again, the higher the AoD, the better the result.

$$AoD = \frac{\sum\{P_{X,Y}|P_{X,Y} > p, X \in R, Y \in E\}}{|\{O_{X,Y}|P_{X,Y} > p, X \in R, Y \in E\}|} \tag{13}$$

4.2   Datasets

To test the accuracy of stream monitoring, we assemble synthetic datasets based on UCR time series archive [18]. We use this archive because these datasets are labeled, and we can precisely calculate the accuracy. More specifically, we concatenate each time series sequence from a training dataset together, but each time series pattern is separated by normalized random walk data with twice the pattern's length to simulate real-world applications. For example, in CBF dataset [18], each training data has 128 data points. We concatenate random

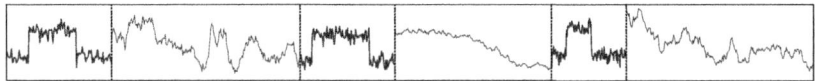

Fig. 3. Example of the data stream built from the CBF classification dataset

walk sequence of 256 data points to connect each and every sequence in the training data, as shown in Figure 3.

4.3   Experiments

All experiments are implemented in Java, and run on Linux Redhat 6.2 with Intel Xeon 3.2 GHz and 2 GB main memory. We test our ASM algorithm using non-overlapping top-k query on ten classification datasets from the UCR time series data mining archive. The k value is set to be the number of patterns within the data stream. We use a concatenated training dataset as a data stream, and use a test dataset as query sequences for evaluation. We then report AoR and AoD of our proposed method comparing with SPRING algorithm, as shown in Table 3.

Table 3. Our experiment result outperforms SPRING in both AoR and AoD

Dataset	AoR		AoD		ASM Parameters	
	SPRING	ASM	SPRING	ASM	$[f_{min} : f_{max}]$	Global Constraint
Synthetic Control	73.82%	74.64%	58.33%	76.85%	[0.6:1.4]	12%
Gun Point	44.31%	53.21%	50.90%	80.11%	[0.7:1.3]	0%
CBF	74.03%	78.10%	60.65%	74.76%	[0.7:1.3]	14%
Trace	72.84%	77.99%	33.21%	76.41%	[0.7:1.3]	50%
Face Four	57.23%	63.15%	55.77%	83.81%	[0.8:1.2]	0%
Lighting 7	47.57%	52.16%	35.41%	89.61%	[0.8:1.2]	0%
ECG 200	60.66%	64.94%	62.87%	88.26%	[0.9:1.1]	0%
Beef	33.89%	38.33%	36.11%	88.94%	[0.9:1.1]	24%
Coffee	79.59%	80.10%	49.23%	94.51%	[0.9:1.1]	2%
Olive Oil	69.76%	72.59%	72.59%	89.69%	[1:1]	0%

4.4   Discussion

From our experiment, we can see that ASM achieves higher accuracies both in terms of Accuracy-on-Retrieval (AoR) and Accuracy-on-Detection (AoD) comparing with the best similarity subsequence matching on data stream under DTW distance, SPRING, as expected. Since SPRING finds minimum distance from all possible subsequences of a data stream, it tries to match the query with a subsequence that gives minimum distance. Therefore, shorter subsequences are preferred. If we have different classes of two patterns as shown in Figure 1, SPRING cannot distinguish. It still reports wrong subsequences, although the size of a query sequence and a retrieved subsequence greatly differ. ASM has great flexibility on limiting the unwanted warping path. Scaling range is also needed since patterns in streaming data are unpredictable. Additionally, S-A band makes a global constraint support sequences of different length.

5   Conclusion

In this work, we have illustrated that the current subsequence matching algorithm (SPRING) on data stream under time warping distance is somewhat inaccurate, and may give undesired results. We then introduce a novel Accurate Subsequence Matching algorithm that has SPRING algorithm as its special case. With the use of a global constraint and scaling range, our experiments have shown to improve the retrieval accuracy on every dataset by a wide margin, while being able to maintain both time and space complexities of $O(n)$.

Acknowledgement

This research is partially supported by the Thailand Research Fund given through the Royal Golden Jubilee Ph.D. Program (PHD/0141/2549 to V. Niennattrakul).

References

1. Ratanamahatana, C.A., Keogh, E.J.: Making time-series classification more accurate using learned constraints. In:Proceedings of 4th SIAM International Conference on Data Mining (SDM 2004), Lake Buena Vista, Florida, USA, April 22-24, pp. 11–22 (2004)
2. Ding, H., Trajcevski, G., Scheuermann, P., Wang, X., Keogh, E.: Querying and mining of time series data: Experimental comparison of representations and distance measures. In: Proceedings of 34th International Conference on Very Large Data Bases (VLDB 2008), Auckland, New Zealand, August 23 - 28 (2008)
3. Sakurai, Y., Faloutsos, C., Yamamuro, M.: Stream monitoring under the time warping distance. In: Proceedings of IEEE 23rd International Conference on Data Engineering (ICDE 2007), Istanbul, Turkey, April 15-20, pp. 1046–1055 (2007)
4. Keogh, E., Ratanamahatana, C.A.: Exact indexing of dynamic time warping. Knowledge and Information Systems 7(3), 358–386 (2005)
5. Kim, S.W., Park, S., Chu, W.W.: An index-based approach for similarity search supporting time warping in large sequence databases. In: Proceedings of the 17th International Conference on Data Engineering (ICDE 2001), Heidelberg, Germany, April 2-6, pp. 607–614 (2001)
6. Yi, B.K., Jagadish, H.V., Faloutsos, C.: Efficient retrieval of similar time sequences under time warping. In: Proceedings of 14th International Conference on Data Engineering (ICDE 1998), Orlando, FL, USA, February 23-27, pp. 201–208 (1998)
7. Zhu, Y., Shasha, D.: Warping indexes with envelope transforms for query by humming. In: Proceedings of the 2003 ACM SIGMOD International Conference on Management of Data (SIGMOD 2003), San Diego, CA, USA, June 9-12, pp. 181–192 (2003)

8. Sakurai, Y., Yoshikawa, M., Faloutsos, C.: FTW: Fast similarity search under the time warping distance. In: Proceedings of 24th ACM SIGACT-SIGMOD-SIGART Symposium on Principles of Database Systems, Baltimore, ML, USA, June 13-15, pp. 326–337 (2005)
9. Zhu, Y., Shasha, D.: Statstream: Statistical monitoring of thousands of data streams in real time. In: Proceedings of 28th International Conference on Very Large Data Bases (VLDB 2002), Hong Kong, China, August 20-23, pp. 358–369 (2002)
10. Wei, L., Keogh, E.J., Herle, H.V., Mafra-Neto, A.: Atomic wedgie: Efficient query filtering for streaming times series. In: Proceedings of the 5th IEEE International Conference on Data Mining (ICDM 2005), Houston, TX, USA, November 27-30, pp. 490–497 (2005)
11. Athitsos, V., Papapetrou, P., Potamias, M., Kollios, G., Gunopulos, D.: Approximate embedding-based subsequence matching of time series. In: Proceedings of the ACM SIGMOD International Conference on Management of Data (SIGMOD 2008), Vancouver, BC, Canada, June 10-12, pp. 365–378 (2008)
12. Yueguo, C., Shouxu, J., Beng Chin, O., Tung, A.K.H.: Querying complex spatio-temporal sequences in human motion databases. In: Proceedings of IEEE 24th International Conference on Data Engineering (ICDE 2008), Cancún, México, April 7-12, pp. 90–99 (2008)
13. Zou, P., Su, L., Jia, Y., Han, W., Yang, S.: Fast similarity matching on data stream with noise. In: Proceedings of the 24th International Conference on Data Engineering Workshops (ICDEW 2008), Cancún, México, April 7-12, pp. 194–199 (2008)
14. Berndt, D.J., Clifford, J.: Using dynamic time warping to find patterns in time series. In: The 1994 AAAI Workshop on Knowledge Discovery in Databases, Seattle, Washington, July 1994, pp. 359–370 (1994)
15. Ratanamahatana, C.A., Keogh, E.J.: Three myths about dynamic time warping data mining. In: Proceedings of 2005 SIAM International Data Mining Conference (SDM 2005), Newport Beach, CL, USA, April 21-23, pp. 506–510 (2005)
16. Sakoe, H., Chiba, S.: Dynamic programming algorithm optimization for spoken word recognition. IEEE Transactions on Acoustics, Speech, and Signal Processing 26(1), 43–49 (1978)
17. Itakura, F.: Minimum prediction residual principle applied to speech recognition. IEEE Transactions on Acoustics, Speech, and Signal Processing 23(1), 67–72 (1975)
18. Keogh, E., Xi, X., Wei, L., Ratanamahatana, C.A.: UCR time series classification/clustering page, http://www.cs.ucr.edu/~eamonn/time_series_data

Author Index

Abramovich, Sharon   99

Balcázar, José L.   76
Buchris, Altina   99

Chawla, Nitesh V.   53
Choudhary, Alok   118

De Marchi, Fabien   1
Dendamrongvit, Sareewan   40

Flouvat, Frédéric   1

Harari, Guy   99
Hsu, Kuo-Wei   28

Koprinska, Irena   106
Kubat, Miroslav   40

Lichtenwalter, Ryan N.   53
Linhart, Chaim   99
Lin, Simon   118

Misra, Sanchit   118

Narayanan, Ramanathan   118
Nattee, Cholwich   141
Niennattrakul, Vit   156

Okumura, Manabu   141

Petit, Jean-Marc   1
Poon, Josiah   16
Porkaew, Peerachet   130

Ratanamahatana, Chotirat Ann   156

Srivastava, Jaideep   28
Supnithi, Thepchai   130, 141

Theeramunkong, Thanaruk   141

Viriyayudhakorn, Kobkrit   141

Wanichsan, Dechawut   156
Weng, Cheng G.   16

Batch number: 09478804

Printed by Printforce, the Netherlands